Rabbit Colonies: Lessons Learned

By Kathryn Kerby

ISBN: 978-1732469716

Published by:
Farm and Ranch Success Publications
a branch of
Frog Chorus Farm, Snohomish, WA
www.frogchorusfarm.com

With help from
IngramSpark Print on Demand Services
www.ingramspark.com

and
Zamzar File Conversion Services
www.zamzar.com

ACKNOWLEDGEMENTS
This guide would have been impossible without the ongoing support and guidance of so many people. First to my family, friends and my husband, who continue to encourage me to go climb that next big mountain. To all my mentors, particularly my holistic veterinarian Dr. Douglas R. Yearout of All Animal and Bird Clinic, Marysville, Washington, for teaching me how to work with what might be the world's most reluctant patients. To all the staff members of the Washington State University/Snohomish County Extension office and Snohomish County Conservation District, for answering 1001 questions over the years about livestock management, animal waste and compost handling systems, water pollution control and runoff mitigation. A big thanks to the Snohomish Health District for ideas on rodent control and disease prevention and for working with us when our rat issue got away from us. And finally, a big fond thank you to all the rabbits, wild and domestic, who have hopped in and out of our lives since childhood. Their gentleness, their cleverness, their quiet persistence, their character and their intense social bonds are all treasures. We are enriched by their presence.

DISCLAIMER
This guide is intended to provide ideas for how to manage rabbits in a colony setting, how to graze rabbits on pasture, and how to build pens which safely allow for both. It is not an engineering manual, and the designs, approaches, methods and plans discussed here are described and intended as suggestions only. It is also not an exhaustive treatise on rabbit nutrition, even though it does reference a variety of authoritative sources for same. The author makes no guarantee or warrantee of any kind, that any specific piece of information will be applicable to any particular reader's unique circumstances. Furthermore, the author strongly encourages every reader to verify that any design, plan or method is in full compliance with local, state, provincial and/or federal laws as may be relevant. Finally, the

author strongly believes that readers must do their own homework to ensure that their planned animal shelters will serve their needs. In other words, don't be an idiot and do something stupid and then try to blame the book for that error in judgment. When in doubt, consult with local experts and authorities. The relevant county Extension office, Conservation District, NRCS office and/or the agricultural department of the nearest land grant university are excellent resources for what works locally. Google works pretty darn well too.

Books from the same author:
The Chicken Coop Manual, 2014

Copyright © 2017, Kathryn Kerby
All rights reserved. Without limiting rights under the copyright reserved above, no part of this publication may be reproduced, stored, introduced into a retrieval system, distributed or transmitted in any form or by any means, including without limitation photocopying, recording, or other electronic or mechanical methods, without the prior written permission of the publisher, except in the case of brief quotations embodied in critical reviews and certain other noncommercial uses permitted by copyright law. The scanning, uploading, and/or distribution of this document via the internet or via any other means without the permission of the publisher is illegal and is punishable by law. Please purchase only authorized editions and do not participate in or encourage electronic piracy of copyrightable materials. For permission requests, please contact the author via the website www.frogchorusfarm.com.

Table of Contents

INTRODUCTION 6

SECTION 1: COLONY RABBIT NUTRITION AND DIET 11
    INTRODUCTION 11
    STANDARD COMMERCIAL PELLET DIETS 12
    ADDING HAY TO THE RABBITS' DIET 14
    TYPES OF HAY 16
    HAY CUTTING TYPES AND CHARACTERISTICS 19
    MORE INFORMATION ON SELECTING DIFFERENT HAY TYPES, AND FEEDING HAY TO RABBITS: 22
    FEEDING LAWN GRASS TO RABBITS 22
    HOW TO FEED HAY OR GRASS WITHOUT WASTE 26
    ADDING PASTURE TO RABBIT DIETS 26
    RABBIT NUTRITIONAL PROGRAM BALANCING 32
    GENERAL DIETARY GUIDELINES 33
    USE OF GARDEN PRODUCE 38
    RELATIONSHIP BETWEEN DIET AND HEALTH, REPRODUCTION, PROFITABILITY/COST EFFECTIVENESS 39

SECTION 2: COLONY RABBIT HEALTH, BEHAVIOR AND PRODUCTIVITY 45
    BREEDING, PREGNANCY AND KINDLING ISSUES 51
    SEXING, POPULATION MANAGEMENT AND HERD GROUPING 55
    IMPROVING HERD GENETICS OVER MULTIPLE GENERATIONS 60

SECTION 3: COLONY RABBIT ENCLOSURE DESIGNS, FUNCTIONS AND MANAGEMENT 62
    INTRODUCTION 62
    PEN MATERIALS 62
    PEN DESIGN 70
    PROS AND CONS OF PORTABLE VERSUS STATIONARY PENS 77
    ONE DESIGN OPTION: SEASONAL PENS 83
    SECURITY AND RABBIT ESCAPE ISSUES 84
    OUR OBSERVATIONS ON COMMON ESCAPE METHODS AND SITUATIONS: ..84
    WHEN RABBITS ESCAPE: WHAT TO DO NEXT 87

SECTION 4: COLONY RABBIT ENVIRONMENTAL ISSUES 91
    PREDATOR ISSUES 91
    PEST ISSUES 95
    PASTURE MANAGEMENT 97
    FODDER GROWING SYSTEMS 104
    WEATHER ISSUES 106

SECTION 5: COLONY RABBIT BUSINESS AND REGULATORY ISSUES 110
    COST EFFECTIVENESS 110
    REGULATIONS 116
    RABBIT OWNERSHIP, POPULATION DENSITY, WELFARE AND SANITATION ISSUES 116
    OUR VISIT FROM ANIMAL CONTROL IN 2011 119
    SALE OF ANIMAL PRODUCTS 127
    RELATIONS WITH NEIGHBORS 128

CONCLUSION 131

PHOTO GALLERY 133

ABOUT THE AUTHOR 141

Introduction

Rabbits were our first form of livestock after moving to our farm property in early 2000. We knew precious little about rabbits prior to buying our first four: a breeding pair of white New Zealands, and a breeding pair of Californians. We had read a number of books, and we had both worked with solitary rabbits as pets when we were growing up. However, we were definitely in for an education when we started to raise them for meat.

Fast forward two years, and our rabbit herd had expanded to having about a dozen breeding does, the same two bucks, and a small clientele who had quite a demand for young rabbits. Our experience with managing the herd had grown over the years, but with the passage of that time we had grown less and less satisfied with conventional management techniques. We very much wanted to manage our rabbits in a colony setting so that they could partake in the various social behaviors we had seen in pet rabbits. We also wanted to get them off the wire and onto grass if at all possible. Thus our experiments with colony management and pasture pens began.

Fast forward another few years, and we had more experience, yet we felt we'd really only started to scratch the surface for what was possible with either colony management and/or pasture-based production. Our experiments with both approaches, whether done one at a time or simultaneously, had given us enough positive feedback that we kept going. Yet we'd also had a variety of "oops" moments, hard-learned lessons, and more than a few mistakes. Surely if we kept at it, we'd work out the kinks.

Then in 2007, disaster hit. Almost our entire herd of nearly 100 rabbits, with multiple breeding pens, quite a few litters at various ages, and a lot of our plans were completely wiped out by the first round of tularemia. This disease, also known as Rabbit Fever, is apparently quite common in our area. The veterinarian who helped us diagnose the issue was amazed we'd gotten as far along as we had without having seen it already. Nevertheless, it was a huge setback. We lost most of the herd in a matter of weeks. I still look back on that time as being like something out of a cheap horror flick – going down to the barn and seeing pen after pen of dead and dying rabbits, which had been up and around and eating and drinking like normal, only a day or two prior. All the experience and knowledge we thought we had, added up to a whole lot of nuthin' at that moment. We were totally unprepared for something like this.

The disease had already swept through our herd and killed off everyone it was going to kill off, before we finally got a grasp on what was happening. Afterwards, we spent quite a bit of time, weeks if I recall correctly, debating whether to continue with the rabbit operation at all. We still had a handful of animals, who either had somehow managed to avoid being infected or had somehow managed to mount a resistance to it (neither of which seemed likely, yet they were still living and breathing). After what seemed like years of limbo, but which was only a few months, we decided that we didn't want to give up what we'd been building. Yes it had been a major setback, but we could either learn from it or give up all the other experience we'd accumulated along the way. The latter seemed a complete waste, so we dove into the former. We read everything we could get our hands on about tularemia: what causes it, how it moves through a rabbit population, how it impacts the body, how to prevent it, and how to treat it.

Fast forward another few years. We had built up our rabbit population again, not quite as large as before, but enough to serve our clientele. And then it happened again – tularemia took down two breeding pens seemingly overnight. But this time we knew what we were looking at, and we were able to dive into action and save the rest of the herd, through a combination of hyper-vigilant hygiene and proactive health measures for those rabbits who had not yet been infected. We came through that episode with most of our rabbits still with us. Apparently, tularemia was a storm that could be weathered with the correct management techniques.

Shortly afterwards, we started renting pasture land on a nearby property where we hoped to graze the rabbits. We had done limited grazing on our home property, and never with very good results. We had so much shade here, and so much hilly ground, that most of our grazing experiments were really glorified exercises in frustration. We thought that rental pasture would be perfect for our needs. While the rental land was nearly ideal grazing ground, nice and flat with lush grass growing from March through November, we encountered a new wave of setbacks. That property was also flush with predators, mostly neighborhood dogs which were allowed to run loose, but also coyotes and lots of raptors. Vermin weren't quite so common there, but weather played a bigger role. That property, while only 10 miles away, required that we get in our truck to drive down there to check on the animals. If a storm suddenly blew up without warning, we sometimes didn't have the time to go put up shelters. Or the winds would rake those exposed pastures and blow away the shelters we did put up. It was another summer spent learning all the things we didn't know that we didn't know.

Fast forward a few more years. We had brought the rabbits home with plans to keep them here for the foreseeable future. The benefits of rental pasture just didn't quite offset the burdens of managing a prey animal 10 miles away. So we shifted gears. We knew that we wanted more garden space, and that we needed to do some bed renovation on the garden beds we already had. The notion that we could use the rabbit pens for that purpose at first seemed rather traitorous, but we decided to at least do some experimental pens with that goal in mind. I don't know if it was the accumulated experience of all those past attempts, or finally finding where rabbits "plugged into" our operation the most efficiently, or if the very stars overhead finally aligned in some benevolent way. But our test beds with rabbits either clearing new garden beds for us, or renovating and fertilizing old beds, worked out really, really well.

We had pens that worked well, we'd figured out shelters that were easy to put up, easy to work with and could take whatever our summer or winter worst could throw at them. Our garden soils were really improving, and our rabbits were healthy. The final cue from the Universe that we were onto something, came when tularemia swept through yet again last summer. Yet this time, it wasn't the devastating illness we'd seen in the past. We had two animals out of the entire herd succumb to the disease, and all the rest of the herd took it in stride. Somehow, that made us feel like we finally knew what the heck we were doing with rabbits.

Finally, let's advance the clock one more time to winter 2016/2017. I was idly reading through some livestock management information, and saw a reference on raising rabbits on pasture. I wanted to see what the latest and greatest developments were in that regard, so I followed the reference to read more about it. While the referenced work was quite good, I noticed a number of comments about "having to learn the hard way" a number of lessons which we'd already learned years prior. In other words, this individual had gone through many of the same mistakes we had. It felt a lot like reinventing the wheel. Was it time to write up what we'd learned, and share it, so that other interested folks didn't have to go through all those same mistakes we made? I started to toy with the idea of writing up some of what we'd learned. The observations and lessons and techniques just started to just pour out of my fingers into the keyboard almost without any effort on my part. Having written professionally for a good chunk of my adult life, I had learned a long time ago that such effortless writing usually meant that an article or a book or a story was ready to be told. Apparently, this one was ready too. And here we are.

On the following pages, you'll see almost the entire sum total of what we've learned since bringing home our first four rabbits in 2000. Yes, we made as many mistakes as anyone could possibly make in that time. Sometimes we made those mistakes more than once. But we always tried to figure out how to do things better along the way. We occasionally had to try two or three (or more) ways to do something before we finally got it right, but we apparently got enough of it right that we like what we have now. Not to say that we've stopped learning; far from it. Even now we're experimenting with new ideas and new equipment and new techniques. There's a quote that comes to mind, from the movie trilogy <u>The Lord of the Rings</u>, where the wizard Gandalf exclaims that you can learn everything there is to know about hobbits in a month, yet after 100 years they'll still surprise you. Rabbits are a lot the same way. We're still learning. For now, however, we'd like to share what we've learned so far. We hope something here is of use to you, whatever your goals may be.
Kathryn Kerby
Frog Chorus Farm
Snohomish, WA
March 31, 2017

Section 1: Colony Rabbit Nutrition and Diet

Introduction
Why include a section on rabbit nutrition, when this is supposed to be about rabbit colony management? For two very simple reasons:
1. One of the most common arguments against managing rabbits in a colony environment, is a general concern about health problems which may stem from that arrangement. In our experience, nutrition and diet are at the heart of most (not all) rabbit herd health issues. Poor nutrition will very reliably result in an increase in infectious disease, parasitism, poor reproduction performance, and poor rates of gain. Yet these issues are consistently held up as examples of problems from colony rabbit systems. I wanted to address that topic head-on and show how any rabbit, regardless of management system, will have health issues if nutrition is lacking. Furthermore, any well-nourished rabbit will be able to survive and thrive in a variety of management systems, including colonies.

2. Colony management is a very natural, almost compulsory component of raising rabbits on pasture. And raising any type of livestock on pasture has several components which all must be balanced, for that interactive system to do well. For most grazing and browsing domestic species, there is plenty of information about rotational grazing, pasture management, forage nutrition, etc. Yet that kind of

information is relatively scarce for rabbits. Furthermore, many rabbit owners are in suburban or even center-city environments, very far removed from any sort of pastoral setting. Yet these owners, and their rabbits, can also reap the many benefits that come from adding fresh or dried grass to the diet, and grazing/browsing activities to the daily schedule, if done correctly.

So without further ado, let's get started.

Rabbit diets vary widely amongst pet owners, hobbyists, small scale commercial operations and larger businesses. The main determinants seem to be feed-time efficiency, feed cost effectiveness (i.e., quality of feed ingredients versus the price/effort/time paid for those ingredients), concerns over the quality of meat, and/or the quality of life for the rabbits. Any given owner may end up exploring one or more of the many feed options available for rabbit colonies, and settle on an approach which is vastly different than another seemingly identical operation.

Standard Commercial Pellet Diets
Let's start the discussion and comparison with what can be considered the "standard" diet of dried alfalfa-based pellets. For owners who have precious little time to spend managing their colonies, nothing beats dry pellets in terms of efficiency. They provide a complete feed for all stages of rabbit life, and the owner doesn't have to fret about balancing nutrients. The rabbits can be fed in a number of individual feeders scattered around their enclosures, or at one or more group feeders. With that approach, the nutritional needs of any given rabbit are met by adjusting

the volume of the pellets offered. The main complications with that arrangement are when a dominant rabbit will drive off other rabbits from feeding, and conversely when a subordinate or mildly ill rabbit will reduce pellet intake or stop eating entirely. Any of those situations can result in a mix of overweight and underweight rabbits within any given colony.

One way to solve all the above scenarios simultaneously is to offer quite a bit more feeder space than what would normally be offered. For instance, one 12" group feeder could easily hold the actual volume of pellets needed for several rabbits in a given rabbit colony. However, any aggressive rabbit could also guard that one feeder and deprive the others of normal intake. The lower on the totem pole any given rabbit is, the less feed that rabbit gets. So offering two or three group feeders will help ensure that each rabbit has easy access to sufficient intake. Similarly, locating feeders in multiple places around the enclosure can help ensure that no one boss rabbit can guard them all simultaneously.

The above is just about the simplest possible feeding system for a rabbit colony, and is fairly typical for those rabbit colonies which are indoors (for instance, housed within a garage, a sunroom, a greenhouse or a barn stall). However, more and more owners are supplementing or replacing alfalfa pellets with a wider diversity of foods for a variety of reasons. That's when things get complicated.

One very common complication occurs when the rabbit colony has seasonal, sporadic, regular or even constant access to pasture, grass or fodder of some kind. Let's split our conversations into two scenarios: rabbits with access to hay, versus rabbits with access to living growing grasses, forbs or other forage.

Adding hay to the rabbits' diet
Let's start with hay. One very common feeding setup for rabbits is to offer them the alfalfa-based pellets as their main feed source, and supplement them with hay as a form of roughage. Feeding hay in any form meets several nutritional and behavior needs for the rabbit which pellets alone don't provide. Why? First of all, rabbits are in class of their own in terms of digestion: they don't have rumens like sheep, goats or cows, yet they need to go through a two-stage digestion process in order to fully absorb nutrients. Since they don't have a rumen, their bodies

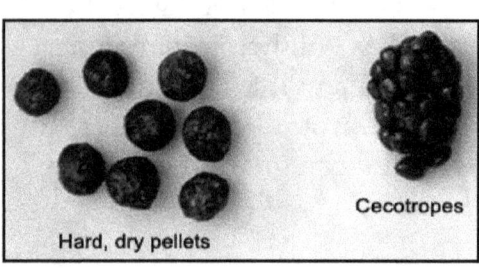

produce two different types of fecal matter. The first is the familiar dry brown round pellet that we see in their enclosures. The second is rarely ever seen, but it looks like a small, glossy mass of dark brown beans, half the size of normal fecal pellets. This mass is composed of different fecal material, called cecotropes. The rabbit's digestion produces them thanks to a different metabolic pathway than the normal digestive process. The cecotropes are usually produced in the early morning hours. The rabbit re-ingests that material (a behavior called cecotrophy) which gives them the chance to re-digest those nutrients. It sounds disgusting, but it is crucial to the rabbit's health. We have occasionally had infirm rabbits recuperating in the

household from either injury or illness, and at first we thought that was a form of loose stool and removed it from the recovery kennel or pen immediately. Then our vet explained their need for it, and we allowed them to consume it as was their preference. Every rabbit which was prevented from eating it, weakened and died. Every rabbit which was allowed to consume it, started to regain strength and vitality. It's that important. As with many other topics we'll touch on, a full exploration of that topic goes beyond the scope of what we can cover here. You can read more about cecotrophy at
http://www.peteducation.com/article.cfm?c=18+1799&aid=2932.

So how does hay figure into that discussion? Fiber is a critical component of the rabbit's digestion system, and pellets alone offer some fiber but not ideal amounts. Without sufficient fiber, a rabbit's digestion will not function as it should. One particular concern is that with rabbit breeds which have standard or long fur (in other words, anything longer than the Rex breed's ultra-short fiber), the rabbit's personal grooming habits will result in a fair amount of ingested fur. That fur can easily be cleaned out of their digestive tract by a diet high in fiber. A diet low in fiber, however, can result in constipation for the rabbit, which can be fatal. They can't expel their normal pellets, and they can't produce their specialized manures for coprophagy. In other words, constipation in general, and hairballs in particular, can kill a rabbit. And that's a really crummy way to lose a prime member of the herd. Happily, a combination of pellets plus hay provides a very nice balance between sugars, proteins and fiber such that the rabbit's natural digestion is kept optimal and dangerous blockages don't occur.

A second benefit of hay is the behavioral enrichment aspect. Rabbits evolved to spend a great deal of time grazing and browsing for their food. When we give them pellets, they can eat a day's worth in little more than an hour. That leaves the rabbit with time on its hands (paws) which can then turn to boredom. A bored rabbit becomes a destructive rabbit very quickly. They will start chewing on anything/everything they can reach, or even start showing more aggression towards each other. We had one buck who decided to start chewing on his own tail. He chewed it off completely before we figured out what the problem was (thanks to our wonderful vet for helping us figure that one out!) By giving them hay to munch on throughout the day, their behavioral need for browsing will be very safely satisfied. I should note that offering hay will dramatically reduce their tendency to chew on things, but it will not eliminate it. It also will not eliminate their tendency to sample other plants in their environment. If the colony happens to have access to valuable plants (houseplants indoors, or possibly expensive landscaping outside), do whatever it takes to protect those specimen plants from the rabbits. Or they'll eventually sample it down to nothing.

Types of Hay
The above-mentioned pellets/hay combination works extremely well, with a few caveats. The first caveat: the type of hay, and the volumes fed, can make a huge different in rabbit health and reproductive status. So let's look at the different types and then make some generalizations accordingly. For those readers who are urban/suburban

and feeding lawn grass rather than hay, bear with me. This hay information is applicable to lawn grass too, but lawn grass has a few of its own quirks to keep in mind. I'll address those separately in a moment.

The richest hay, and often the most expensive, are the legume hays such as alfalfa. Clover hay is also a legume, and is available in some areas. These hays are extremely high in both sugars and proteins, being much higher as a rule than any other form of hay regardless of species. Some rich grass hays will come close to poor-quality legume hays, but even then that's not very common. I might get some argument on this next statement, but I'll go ahead and state that feeding alfalfa-based pellets PLUS alfalfa hay is too rich for most adult rabbits. In our experience, the adults will feast on both the pellets and the hay and gain unacceptable amounts of weight. While an overweight buck is not a huge management problem, an overweight doe will stop ovulating. That becomes a problem very quickly amongst rabbit herds that need to maintain a high level of productivity. There are only two instances where I would feed alfalfa pellets plus free choice legume hay. The first is for a doe or group of does who are being bred back very quickly after kindling, such that they have very high nutritional requirements from pregnancy, lactation and/or general body maintenance. The second group I'd feed alfalfa pellets plus alfalfa hay, is a group of fast-growing kits just past weaning.

A second type of hay is known locally as horse hay, i.e., richer hay types intended for horses. Timothy and orchardgrass both fall into this category. These are not quite as rich as the legumes in terms of proteins but they still have very high sugar contents if they were harvested and baled under normal haying conditions. The extra-long leaf blades for either hay type stay easily digestible and relatively free of lignin far longer than other grass species, so the plant can grow much larger and keep a good nutritional profile compared to other grasses. When feeding these hays to rabbits, there is a risk that a normal amount of pellets combined with free-choice hay will still result in overweight rabbits. However, the hay itself doesn't provide quite enough protein on its own to substitute very much for the alfalfa pellets. If this is the type of hay which is most easily available, give the rabbits their full measure of pellets, and give them limited quantities of hay, say a handful per rabbit. They'll benefit from the roughage without gaining weight.

A third type of hay is generally the cheapest to purchase, and might be known by a variety of names, including "cow hay", "local hay", "meadow hay" or "mountain hay", amongst others. In general, it's coarser and better suited to true ruminants. These hays vary widely by region, and are composed of the species common to any given area. In this area, we have reed canarygrass, ryegrass, brome and a few others being offered as common local grass hays. These are generally the lowest in proteins and sugars, and tend to go to lignin earlier in their development so they get stemmy faster. These hays can often be given free-choice when combined with alfalfa pellets, for a very nice blend of high quality nutrition and plenty of fiber. We use reed canarygrass here for a multitude of purposes, including free-choice hay, bedding, nesting material, and mulch.

Hay Cutting Types and Characteristics

Finally, I should give an explanation about hay cuttings. In this context, the term "cutting" is a reference to when the hay was cut, not only in terms of season but also in terms of the plant's growth. It is not an exact term, but rather a thumbnail estimate for what to expect from that particular batch of hay.

In most parts of the country, hay is sold by 1st cutting, 2nd cutting or sometimes a 3rd cutting, regardless of species or type. A 1st cutting, as the name implies, is the first cutting of the year for that particular field, and generally occurs in late spring. A 2nd cutting would generally occur in mid-summer, and a 3rd cutting would be late summer to early fall. So for instance, if your local feed store is advertising both 1st and 2nd cuttings of orchardgrass hay, then the 1st cutting occurred in late spring/early summer and the 2nd cutting is from later in the same year. Some northern parts of the country don't have a long enough growing season to get a 3rd cutting of hay (for instance, we're lucky to get a second cutting here in our northwestern maritime climate). Conversely, southern parts of the country, particularly the southwest, might have cuttings almost monthly for most of the year.

So what difference does the cutting make? In general, the first cutting of hay will be the cheapest, and the coarsest. This is because ideal haying conditions require a long stretch of hot, dry weather, yet grass grows fastest during cool, damp weather. That discrepancy means that grass is at its prime for hay cutting while the weather is still too fickle for successful haymaking. By the time the weather settles down long enough to actually cut, dry and bale the hay, the grass has grown past its prime and has started to turn stemmy and coarse. So 1st cutting hay is a lower grade than subsequent cuttings, and thus cheaper. That's good news for rabbit owners, because generally we want to feed the coarser hays if our rabbits are also on standard pellet diets. However, if we're feeding straight hay to our rabbits without pellets, 1st cutting may not be rich enough.

A 2nd cutting, by contrast, is usually the most expensive hay. This is because mid-summer is when grass growing and hay making actually line up the most frequently. If a grass has already been cut once during the year, say in late May or early June, then it will have regrown to haying length by mid-July. In many parts of the country, mid-July happens to be ideal haying weather, with hot dry conditions and infrequent rain. So the grass is nutritionally at its prime, during prime cutting, drying and baling conditions. That's why 2nd cutting hay is generally the finest grade of hay, and the most expensive. For rabbit owners who want a high-fiber cheap hay, this isn't the cutting to use. Conversely, for rabbit owners who are not feeding pellets, and need the highest grade of hay they can get, this is the stuff.

Some parts of the country (like us) don't get 3rd cuttings, because by the time the grass has regrown from the second cutting, say mid-September, the weather has turned dramatically cooler and rain is more common. As we saw in the section on making hay, cool damp weather is exactly what we don't want while haying. However, this varies by location and by year, with some locations getting a wonderful 3rd cutting some years. Just know that a 3rd cutting can be something of a gamble. The harvest may have been rushed a little, due to weather concerns, so the grass may have been cut shorter than usual, or it may have been baled up wetter than usual. The former will result in a much higher protein and sugar content than even 2nd cutting, but the latter can result in spoiled hay. If you see a 3rd cutting of hay being advertised at a really attractive price, check it out pretty carefully. It might be wonderful hay, or it might be starting to spoil inside. Also consider the location, and what else is available. If your location had a very long, dry summer, and everyone is offering 3rd cuttings, then it's probably OK. If, however, you happen to know that you had a surprisingly short growing season due to early autumn, and only one or two sources have 3rd cuttings available, check those cuttings very carefully before laying out a lot of money for what might seem like a good bargain.

The last thing to consider with all this talk of cuttings, is that hay is now being shipped all over the country, from regions with vastly different weather. So any given feed store might have access to hays grown in much different climates. For instance, we only get one or possibly two cuttings here, but hay is regularly shipped to our regional feed stores from haying regions to the south and east of us, which have very different growing seasons. So we have access to 2nd and 3rd cuttings locally, even though the hay was actually grown hundreds of miles away.

Going into more detail about hay is going to take us too far off topic, so I'll wrap things up here. As with other topics in this discussion, the topic of "which hay is best" is much too large to really cover here. Our best suggestion is to look at what hay you have available locally, determine which of the three categories listed above that hay falls into, then design your feeding program accordingly. If you're really not interested in becoming a hay expert, it doesn't need to get much more exotic than that. Or, if you really do want to learn more about hay types and qualities and nutritional profiles, by all means pursue that interest.

More Information On Selecting Different Hay Types, and Feeding Hay to Rabbits:

http://www.guinealynx.info/hay_chart.html

http://www.hobbyfarms.com/all-hay-is-not-equal-choose-your-livestocks-carefully/

http://myhouserabbit.com/rabbit-care/hay-for-rabbits-the-basis-for-a-healthy-diet/

http://www.therabbithouse.com/diet/grass-hay.asp

https://www.sandiegorabbits.org/diet.php

Feeding Lawn Grass to Rabbits
Now let's talk about lawn grass. Lawn grass, as mentioned above, might be the closest that urban and suburban residents ever get to working with hay. The good news is that yes, lawn grass can be used. Additionally, many of the principles we talked about with hay apply to grass too. The details, however, can get a little tricky compared to hay, so let's address those now.

Lawn grass, nutritionally speaking, is generally about on a par with rich grass hay, almost bordering the legume hays in terms of both sugars and proteins. It also has the highest moisture content of any category of grass production, because it is so young and leafy. It has virtually no stem since it is generally 6" or less in height.

However, standard methods for harvesting lawn grass (i.e., the conventional lawn mower), is vastly different from standard hay production. First of all, many lawn mowers basically shred or bludgeon the individual grass blades, rather than cut them. That results in a very ragged end on each individual grass blade, which allows for extremely rapid moisture loss from within the grass blade itself. Secondly, the grass is typically gathered into a bag with very little air circulation. Third, lawn clippings are often emptied from the mower bag into either a waterproof yard waste container, or an even more airtight trash bag. Fourth, lawns are generally mown in mid-late spring and early summer, with high humidity and sometimes high heat. Those details very quickly convert an otherwise nutritious supplement into a soggy, fermenting, potentially poisonous mess.

Suppose you'd like to harvest your lawn as a grass supplement for your rabbits. It can be done, but it needs to be done a little differently than simply mowing the lawn. So let's step through that scenario again and look at what can be done differently. Standard lawnmower blades will shred the grass ends, such that the grass loses moisture very quickly. That by itself is actually a good thing; drying grass down very quickly will preserve the maximum amount of nutrients. It's the next step where things need to change. Normally, it is then gathered into a lawnmower bag where it heats up and mats down. That moisture doesn't have anywhere to go, so it starts to coat the outside of each grass blade. Some of that moisture will contain trace amounts of sugars. Combine the airtight bag, the moisture, the sugars and the average heat which is typical of lawnmowing season and you have a perfect recipe for low-oxygen fermentation. That will quickly result in a bloom of alcoholic byproducts, fungal blooms and other microbial nasties which our rabbits really don't need.   Toss the now-soggy mess of grass clippings into a trash bag and let it sit for even 12 hours, and that process just accelerates. The longer it sits, the more the fermentation process gets really cooking. If you've ever noticed how lawn clippings heat up after a day or two, that's why.

Can we modify this process to create edible homegrown hay? Absolutely, but it requires a lawnmowing procedural change, and perhaps a bit of creativity. The first thing we need to change is the bagging of clippings during mowing. If we know that we want to use those clippings as hay, the single best thing we can do is leave them on the freshly mown lawn for a day or two. That exposure to air and light will result in a very fast dry-down time. After 12-24 hours (sometimes even less in dry climates) you can go back and look at the freshly cut lawn, and see how the individual cut blades have shrunken down and dried out. Voila! You have just created hay. To answer a question which might already have surfaced, isn't leaving grass clippings on the lawn, bad for the lawn? Only if a) the grass blades are relatively long, and b) if they're allowed to mound up. So-called "thatch" which most people abhor is actually a mulch of several layers of miniature hay. But lawn lovers, never fear. We're not going to leave it there.

The next step is to wait until after morning dew has dried off, and then rake up the hay into a paper bag or cloth sack (an old pillow case works well for small amounts), and then put that hay in a dry shady location. Feed it out over time and your rabbits will be delighted.

What about if your lawnmower has a bagging system which isn't easy to disconnect? Or perhaps you have a spouse or partner who insists that the grass clippings must be bagged? In that instance, some creative intervention can still save the day (and the grass). Even if the bagger is hard to remove from the mower, the bag will be designed to allow for easy removal of the grass. In that instance, pull the fresh-cut grass out of the bagger as quickly as humanly possible. Secondly, that collection of still-wet grass will need a place to dry out with plenty of air movement both above and below it.

One possibility is to set up a drying table, ideally made of mesh, and then spread the lawn clippings out on that table. It's hard to say in advance how big that table would need to be; a small urban yard might only need a picnic table sized area, while a substantial suburban lawn could need several such tables and still not have room for an entire lawnmowing session. Use your best judgement, experiment with different areas and just see how it goes. You want the grass clippings to be no more than about 1" in depth, and it's best to turn the clippings over once or twice as they dry so that the lower levels get good air exposure. If you can create a true mesh platform, turning won't be necessary. If space is at a premium and you're already trying to work with a less-than-ideal setup, turning the grass frequently is the key. You'll know it's done when the grass is not quite crunchy, but is definitely fluffy. If it still has weight to it, let it dry a little more. Finally, make sure the table is provided with shelter during rain. Letting the grass get wet again not only sets you back to where you started, but a bit more of the nutrients wash out. If the grass does get wet but is successfully dried afterwards, it can still be used. Just know that it will have a nutrient profile more closely related to the so-called cow grass hay we discussed above, rather than a rich grass hay.

How To Feed Hay Or Grass Without Waste
Regardless of which hay (or grass) is offered, one excellent way to get the most value from it is to offer it in hay racks or semi-protected stands, rather than simply tossing it on the ground. The rabbits will spend hours pulling individual blades of hay from a vertical hay feeder. Contrast that to spreading it on the ground, where they'd simply trample it and then never eat it. If your hay is cheap enough that you don't mind a big percentage of it going to bedding (we figure 50% of hay on the ground ends up as bedding), then that's fine. If hay is an unwanted expense and you want to make the very best use of it, any type of homemade hay rack works extremely well. We've used 2"x2" woven fencing material to make little feeders of roughly 4" wide by 18" tall by 2' long, which work really well for rabbit groups. They're lightweight enough we can simply hang them from a convenient branch, or build a little stand for them. That size rack will hold two flakes, which will feed a group of six rabbits for a few days.

Adding Pasture to Rabbit Diets
Now let's make things a tad more complicated and talk about living grasses, i.e. pasture. While a comprehensive

analysis or description of pasture (or lawn) forage quality/quantity goes well beyond the scope of what we're trying to cover here, we can make some generalizations. We know that grass will be highest in proteins, sugars and water content when it is young and growing fast (say, anywhere between 2" and 10" tall). That is when it's at its most palatable for most species. Conversely, grass which has grown rank and tall will have proportionally less sugars, less protein and more indigestible lignin (the woody component of stems). Most commercially available hay tries to split the difference between those two extremes, and features grass species which may have long stems (for volume at harvest and fiber content) combined with long blades of grass (to provide the high proteins/sugars content).

A pasture (or lawn) can be at any stage of growth at any given time, but some stages are much more suitable for rabbits than others. Additionally, different classes/ages of rabbits would do best on different stages of pasture growth, thanks to the rabbits' differing nutritional needs. For instance, we have occasionally managed bachelor bands of rabbit bucks in a group, well away from reproductive does. Those bucks have already reached mature size and weight, and they're not breeding. So their nutritional needs are relatively low. Furthermore, they don't have the social obligations of breeding a group of females, so they are tending towards the "inactive/bored" end of the scale. By putting them in a pasture of over-mature grass, they get very busy creating tunnels and hideouts, and they can eat as much of the over-mature, low-nutrient quality grass as they like without getting overweight. The tall grass itself encourages them to move around more than they might have if exposed on a field with shorter grass. So that particular combination of old grass/mature non-breeding rabbits can work well.

Now contrast that with a litter of weaned rabbits, all the same age, out on pasture. They are growing at a remarkable pace, and they need all the nutrition that they can get. If turned out on an over-mature pasture as described above, they would be very active and create the same tunnels and burrows, so they would never be bored. However, that grass, as a primary feed, would never support their growth rates. Their growth rates would slow down and stop, they'd lose weight, and they'd eventually die (perhaps in as little as a few weeks) if turned out on a pasture of over-mature grass. However, take that same bunch and put them on a well-managed pasture of fast-growing grass, and they would do very well as long as the grass has high-enough protein and sugars for their rapid metabolism. This is how the Joel Salatin's Polyface Farm grows out rabbits for market in their "rabbit tractor" settings.

A quick note about the Salatin setup. As anyone who has read Joel's books knows, his parents bought that property as a woefully run-down property, which had been farmed for decades without much concern for maintaining fertility or pasture health. Over the passage of time, and with a lot of elbow grease, Joel's dad improved the pastures a little at a time. Joel has further improved those pastures with his rotations of different livestock species, and his promotion of different plant species. His books talk about his pastures being something of a salad bar (one of his books is entitled <u>Salad Bar Beef</u>). That is a reference to the fact that his pastures have a wide range of plant communities which have thrived under intensive rotational management for the last 60 years. That can easily be held up as an ideal for which we're all striving.

Putting a rabbit tractor out onto those pastures works because the Salatins know from long experience that their grass and forage mixture will provide superior nutritional support for young rabbits. A worn out old hay pasture, or the typical front lawn, will not have the same diversity or nutritional profile. The single biggest fault I've seen in talking with people who want to pasture their rabbits, is the assumption that their grass is "good enough", or even as good as Joel's. Sadly, that is usually not the case. On those lesser pastures, the rabbits might grow OK for a while, but then they'll weaken over time. When I talk to people who have tried and abandoned pastured rabbits, many of those folks talk about how their rabbits got sick and died from a variety of illnesses and/or internal parasites after only a few months on pasture. Those folks will talk about how wonderfully green the pasture was, and blame hygiene issues for those situations. The real cause, however, was malnutrition, which weakened the rabbits such that they would succumb to whatever particular microscopic beasty happened to come along. Those folks rarely if ever took the step of learning how to property manage a pasture, and had no idea what nutritional profile they were working with. Rabbits raised on vibrant, diversified, well-managed pasture have strong enough immune systems that they aren't generally at risk of such issues. But the healthiest rabbit herd in the world can't survive long on poor grass, no matter how green it looks.

So any plans to rely on pasture as a major food source for rabbits, should include a hard look at just how nutritious and well-balanced that pasture really is. A group of rabbits tossed onto a patch of hardscrabble ground and made to fend for themselves, is not going to give the same bountiful production as a group of rabbits which are carefully managed on healthy pasture. Anyone who wants to get started with keeping rabbits on pasture, would do well to first assess that pasture's condition and nutritional profile. Then a complete feed ration can be developed for the rabbits, which will ensure their health while restoring the pasture back to its rich, diversified, thriving potential. The good news is, the rabbits can still benefit from being on a pasture while it heals, as long as their overall dietary needs are met. And the pasture will heal faster with the presence of carefully managed rabbit herds. That symbiotic relationship can be a thing of beauty to behold when each is carefully managed for long-term vitality.

For more information on the Salatins' rabbit operation, check out the blog entry written by one of their interns, listed below. For information on evaluating and restoring pasture productivity (for any species), we have some references for additional reading listed below as well.

This topic of pasture management is vast, with a lot of different tangents and management methods. The good news is, there is a tremendous amount of information available about Management Intensive Grazing (MIG), rotational grazing, mob grazing, strip grazing and a wide variety of other variations on this theme. All of these approaches have pros and cons, and virtually all the printed or online research articles will focus on one of the major livestock species (horse, cow, sheep or goat). Precious little research has been done on pastured rabbits. If you want to turn rabbits out on grass primarily for the fresh air, exercise and sunshine, and the grass is merely a supplemental diet, then almost any pasture will work as long as it's not muddy, infested with toxic weeds or otherwise inhospitable. However, if you want to really "work" those pastures (or lawn) with rabbits as a primary grazing animal, and the pastures serving as a primary nutritional source, then you may need to really study up on as much pasture management and nutritional materials as you can find. Even reading about other species of livestock will help you understand how pastures can be managed. That diversity of information, combined with a good understanding of what rabbits need nutritionally, will serve you very well. To get started on reading about pasture management, check out these resources:

Rabbit Tractors and Rabbit Pasture Management:
http://www.mofga.org/Publications/MaineOrganicFarmerGardener/Winter20092010/Rabbits/tabid/1392/

System for Raising Meat Rabbits On Pasture, Operational Profile:
http://www.sare.org/Learning-Center/From-the-Field/North-Central-SARE-From-the-Field/Julie-Engel-and-her-Sustainable-System-for-Raising-Rabbits-on-Pasture

Pastured Rabbit Cage Development Research Project, Final Report:
https://mysare.sare.org/sare_project/FNE01-354/?page=final

More Information on the Polyface Pastured Rabbit System, written by a Polyface Farm Intern:
http://polyfaceapprentice.blogspot.com/search?q=rabbit

Letting Rabbits Graze on Grass Lawns:
http://www.therabbithouse.com/blog/2009/05/22/rabbits-eat-grass/

Rabbit Nutritional Program Balancing
So perhaps now is the time to shift attention away from pastures for a moment, and look more closely at a rabbit's nutritional needs throughout life. We've spoken in very general terms so far about the fact that rabbits need different nutritional programs at different times/phases of life. Let's look more closely at those needs and how to meet them.

We have three very different sources of information about rabbit nutritional needs and ration balancing. One such source is from the medical research community. Rabbits have been used as an integral part of medical research for many decades, and those labs have worked out very precise guidelines for how to keep rabbits healthy and well nourished. Granted, lab animals are subjected to circumstances and health challenges which most other rabbits will never face. However, a rabbit is a rabbit regardless of why it's being kept, so those nutritional guidelines don't change much.

A second source of information comes from studies of wild rabbit nutrition and grazing/browsing behavior. Here again, wild rabbits will face extremes of temperature, food availability and social/predator pressures which your rabbits may never experience. Yet a wild rabbit's food choices, made while living under those challenging circumstances, can give us valuable insights into ways to ensure our rabbit herds are as healthy as possible.

A third perhaps unconventional source of information comes from the pet rabbit industry. Rabbits are becoming a more and more common house pet, alongside or instead of the more traditional dogs and cats. Granted, the folks who keep rabbits as pets may not think they have much in common with the folks who keep rabbits for meat and pelts, and vice versa. Yet the pet rabbit community has worked out some solid guidelines for feeding rabbits to ensure long, healthy lives. That information is useful no matter why a person keeps rabbits.

General dietary guidelines
So from that standpoint, what can we learn from each group about feeding our rabbits? To borrow an old phrase, let's start at the beginning, and look at wild rabbit diets. This could be considered a baseline or foundation; after all, if wild rabbits are surviving and reproducing with no help from human caretakers, they must be doing something right.

The modern meat and fiber rabbit breeds all trace back to the European Rabbit, *Oryctolagus cuniculus*. This rabbit was originally native to areas surrounding the Mediterranean Sea, such as southern Europe and northern Africa, but over the ages it has been introduced to all the continents except Antarctica. It has become a serious invasive species in some areas, particularly Australia where populations went from

24 introduced rabbits in 1859 to over 6 million wild rabbits less than a century later. It was raised for meat and pelts at least as far back as the Roman Empire.

Rabbits are general herbivores. Wild rabbits will browse a wide variety of grasses, forbs (wide leaved plants), wild grains, roots, tree saplings and leaf buds. Gardeners and farmers alike know that rabbits will readily eat a variety of crops such as brassicas, lettuces, root crops, grains and ornamental shrubs. One factor which is important to both wild and domestic rabbits, is the availability of nearby shelter. Wild rabbits are considered a keystone species, meaning that they are preyed upon by a huge variety of carnivores. A rabbit's eating habits then, are shaped by this omni-present risk of predation. In the wild they prefer at least some shrub or brush cover to provide immediate shelter from predators. That seems to be most important to rabbits during daylight feeding, because they choose not to range very far from such shelter. At night they range further into open areas to graze and browse.

Much of a wild rabbit's diet is generally low in protein, sugars (energy source) or both, and it is often high in indigestible materials such as woody plant materials and bark. As we've already seen, their coprophagy allows them to digest materials twice, which is thought to give them access to trace minerals they would not be able to absorb through a single digestive process.

Now that we know a bit more about wild rabbit nutrition, what does that suggest for captive rabbit nutrition? Here is where laboratory rabbit diets can tell us a lot about how to manage groups of rabbits, very cost effectively, for

maximum productivity and health at minimum reasonable cost. The Merck company is a long-standing member of the medical research community, and they have long history of publishing materials to assist caregivers in both the medical and veterinary fields. The Merck Veterinary Manual is one of the bibles of modern veterinary care for almost all domestic species; your particular vet probably has a worn copy somewhere in his/her library. So it's no big surprise that the Merck website has an extensive set of information on the particular dietary needs for laboratory rabbits. Those pages can be accessed at
http://www.merckvetmanual.com/exotic-and-laboratory-animals/rabbits/nutrition-of-rabbits#v5625687.

Let's note their particular guidelines from that webpage:
- All the components of the basic diet (i.e., protein, fiber, fat, and energy) should be managed in consideration of the life stage (growth, gestation, lactation, maintenance), breed, condition, and lifestyle of the rabbit.

- Ratios should meet the nutrient requirements of the National Research Council (see Table: Nutrient Requirements of Rabbits).

- Pelleted rabbit feeds provide good nutrition at reasonable cost.

- Fresh, clean water should always be available.

- Prolonged intake of typical commercial diets containing alfalfa meal by laboratory or pet rabbits kept for extended periods under maintenance conditions may lead to kidney damage and calcium carbonate deposits in the urinary tract. Reducing the calcium level to 0.4%–0.5% of the diet for nonlactating rabbits helps reduce these problems. This can be accomplished by feeding pelleted diets with a timothy hay base. Ad lib timothy hay is usually recommended for the maintenance diet of adult rabbits.

- Adult pet rabbits not intended for breeding should be fed a high-fiber pelleted diet, restricted to ¼ cup/5 lb. body wt./day to prevent obesity and maintain GI health.

- Rabbits are efficient converters of poorly digestible materials to meat. Therefore, it is easy to overfeed or underfeed does and growing, adolescent rabbits (fryers).

- The amount to feed depends on the age of the fryers or on the stage of pregnancy or lactation of the does.

- A general rule in feeding fryers is to feed all that can be consumed in 20 hr., with the feed hopper empty ~4 hr./day.

- Does are usually fed ad lib once they kindle. The general practice is to bring does from restricted to full feed slowly during the first week of lactation.

- Does bred to kindle five times during the year generally have their feed restricted between litters; those bred intensively should be on full feed continually once they begin the first lactation.

Now that we've seen those recommendations, let's compare that to some recommendations specifically for pet rabbits. The House Rabbit Society is an international advocacy group (based in the USA but with additional offices abroad) which provides information and support for pet rabbit owners. As such, they have done a lot of work to determine what nutritional guidelines work best for a rabbit's well-being, not only in terms of overall health and longevity but also in terms of a rabbit's behavioral needs. Additionally, since many rabbit-loving households have more than one pet rabbit at any given time, they have broad experience in making sure that all rabbits in any given group have suitable access to adequate foods. Finally, they also have experience in helping rabbits achieve and maintain healthy weights after either being malnourished or overfed. That experience can be crucial for rabbit owners who either suspect, or have become convinced, that their rabbit feeding program must be revised to address or avoid a variety of health issues.

The HRS website (http://rabbit.org) has, at the time of this writing, 24 different articles devoted to rabbit diet and nutrition. Their general recommendations for all rabbits are available at http://rabbit.org/what-to-feed-your-rabbit/. In short:
- 80% of the daily diet should take the form of a good quality grass hay such as timothy or orchard grass.

- Alfalfa-based pellets should compose less than 20% of the diet, and ideally should be less than 10% or eliminated altogether (we'll touch on this point later).

- Feed a minimum of 1 cup vegetables for each 4 lbs. of body weight. At least three different vegetables should be fed any given day, with a wide assortment of vegetables chosen over time.

- Small quantities of fruit should also be fed, at a rate of 1-2 tablespoons per 5 lbs. of body weight (none if dieting). High-fiber fruits are best.

I could go on at length about the merits of solid nutrition for rabbits, but that goes beyond the scope of this work. I'd also be reinventing the wheel to a certain extent since that information is already available elsewhere. Specifically, the Merck and HRS websites provide a great deal of additional information about feeding various classes of rabbits. Those websites are a veritable treasure trove of information on diet, and provide wonderful sources for specific stage-of-life information. I would heartily encourage anyone interested in further exploration of this topic, to read up on their available materials.

Use of garden produce
Now that we've started to talk about introducing veggies and fruits into the rabbit diet, the next logical topic is whether to purchase those veggies, or simply grow them. This is a choice which each rabbit owner will need to make for him/herself, based on a lot of variables which go well beyond the scope of this work.

Specifically, some folks already garden and are already in a position to offer their rabbits a wide selection of fruits and veggies, and even edible weeds. Other owners are either not able to garden, or have no wish to, for a variety of reasons. Either situation is fine. The major question to ask is this: what is the overall most satisfying answer for your particular situation? Don't feel pressured to grow your own if you've already got basketful of other responsibilities. Simply offering your hops a variety of store-bought produce will be a vast improvement over not having access to veggies at all. If, on the other hand, you're already producing a variety of veggies thanks to gardening and/or feeding other livestock, then by all means. In either case, try to follow the recommendations already listed above for how much to feed the rabbits based on whatever else they are eating.

Relationship between diet and health, reproduction, profitability/cost effectiveness
So now that we've looked at all the different possible components of a healthy rabbit diet, how do we put it all together in a well-balanced way? The answer will vary according to a number of variables, and we can't make one blanket recommendation. What we have found here, however, is:
1. Start with what has proven to work, and then branch out from there. One of the mistakes we see new rabbit owners make, is that they want to start with rabbits while using a 100% homegrown diet. That's one heckuva learning curve to climb. If a person has no prior experience with raising rabbits, then he/she has no idea what a normal healthy rabbit population looks

like – what the body condition should be, what the reproductive rates should be, what the weaning rates should be, the slaughter weights and carcass composition, etc. Start with healthy rabbits, and feed them a standard diet of pellets with hay as already described. Even if your eventual goal is to feed a 100% homegrown diet, this "intro diet" will give you a baseline against which to gauge the success (or failure) of your homegrown diet results.

2. Once you get some experience, don't get locked into one and only one feeding program. Rather, start (and hopefully continue) to experiment with different approaches and different combinations. Some combinations will work out surprisingly well. Others will present you with a number of unforeseen complications which might make them impractical.

3. Try any feeding experiment on a small scale. One very telling experiment we've run a number of times, is to feed several same-age litters of rabbits with different feeding plans, and see how they compare over time. If the litters are within a few days of age with each other, then a side-by-side comparison of two feeding programs can give you tremendous information in a very short time. Growth rates, health scores, final weights, incidental behaviors such as chewing or aggression or fur-pulling, etc., will all be slightly different between the two feeding plans. That is possibly the single best experiment of all that you can run, because so many of the other variables (age,

weather, season, etc.) are cancelled out by running the two litters side by side on different diets.

4. Remember that different life stages require vastly different nutrient profiles. The above example is wonderful from the perspective of "scientific method experimentation", because you have controlled for many possible variables, and you have multiple test subjects following two different feeding programs. However, also keep in mind that growing litters have the highest possible nutritional requirements of any age group, so whatever experiments you want to run MUST first provide them with that nutrition. That group of rabbits is also often the money-makers, in the sense that you'll be selling them in a matter of weeks, but likely hanging onto the breeders for a longer period of time. If you're not sure that your nutritional program is solid enough yet to meet the needs of a young group of growing rabbits, consider running your feed experiment on two bachelor groups, or perhaps on two groups of older females. You'll still get very good comparison between the two, but you won't be endangering your income in the process.

5. Take good notes, and lots of them. Running any sort of comparison between two groups of animals requires that you actually have solid data when you're done. At the very least, start with the body weights of each rabbit before you start the experiment. After all, how can you tell if they lost weight, gained weight or maintained weight during the program, if you don't

know what their weights were before you started? Also record their lumbar scores, so that you have a good idea of what sort of body condition they were in. Record not only the variety of foods they were fed, but also weights of each per day, and what sort of food it was. For instance, "fed a flake of hay to group B" doesn't give you much information. A statement like "fed a 3lb flake of 1st cutting orchardgrass hay to group B" gives you quite a bit more information.

6. Don't forget to add up the costs for each experiment: the cost of the different ingredients, the time you spent putting things together, and the effort required to do so. Many, many rabbit owners complain about the high cost of rabbit feed, but they have never actually gone through the motions to grow AND harvest AND store AND prep their own home-grown feeds. If they were to ever actually go through those motions, they'd be surprised to find out that their "expensive" purchased feed is actually cheaper in the big picture, than doing it themselves. This is a purely individual issue, and each person will have vastly different circumstances which all contribute to how this scenario pencils out. At the very least, however, document how much each experiment really truly costs, so that any comparison with store-bought feeds (and between different homegrown feeds) will be both qualitative, and quantitative.

7. Don't be surprised when you get different results than you expected. Different results don't mean the

experiment failed; rather, that's solid information for you to use to refine your ideas.

8. Finally, don't be surprised when your own circumstances change, such that a past feeding program which didn't work, might now makes sense. Or conversely, that a feeding program which worked quite well for a long time, just isn't working anymore. There's an old saying that "it's never the same river twice", meaning that every day brings slightly different circumstances to the same general situation. Allow for those changes, and try to roll with them. Being flexible can be the single biggest asset you ever develop in terms of feeding your rabbits well. Or in life, for that matter.

The topic of feeding rabbits could go on indefinitely, with a lot more detail about ingredients, nutritional needs, cost effective sourcing, pasture balancing, hay testing and quite a few others. Frankly, it was difficult to decide where to draw the line on any given topic; I wanted to hit that "Goldilocks" point of providing just the right amount of information. Not too much, and not too little. I have tried to present a balanced introduction to these various topics, with the hope and expectation that any given reader will pursue whichever topic(s) are of great interest. Towards that end, I want to provide some resources which will be useful to someone wanting to learn more about the vast subject of rabbit nutrition, feeds, feeding and feed ration balancing:

Extensive information on wild rabbit natural history, including diet choices, grazing patterns and food preferences:
http://eol.org/pages/327977/details

A fascinating and detailed overview of the domestic rabbit's rise as a domesticated animal during the last 1000 years, including a tally of production improvements over the last 100 years:
http://www.fao.org/docrep/t1690E/t1690e03.htm

The House Rabbit Society home page on rabbit diet and nutrition (all their various dietary articles are listed here):
http://rabbit.org/category/care/diet/

The Merck home page for rabbit nutrition, with links to other resources:
http://www.merckvetmanual.com/exotic-and-laboratory-animals/rabbits/nutrition-of-rabbits#v5625687

Section 2: Colony Rabbit Health, Behavior and Productivity

As with any domestic animal species, the topic of health and behavior is vast, deep and potentially complicated. My goal here is not to provide an exhaustive review of known rabbit health and behavior issues, but rather to introduce those which are particularly relevant to colony housing. I want to specifically address those health or behavior issues which are all too frequently used as reasons or justifications for NOT managing rabbits in a colony. In our experience, those issues are either misunderstood, misrepresented, or simply flat out inaccurate.

Diet: This single topic is quite large, and certainly applies to rabbits as a whole rather to rabbits in colonies. However, I chose to address it earlier in this write-up because it is so important for the long-term health and productivity of rabbits regardless of housing or management systems. I address it again here because so many health issues which are blamed on colony management, are in my opinion actually directly or indirectly caused by poor nutrition. I won't rehash what I already wrote in previous sections, but rather I'll mention in passing a few of the more important points:

1. a well-balanced diet, featuring suitable amounts of proteins, sugars, fiber, vitamins and minerals, is the single best way to keep rabbits happy healthy and productive.

2. No housing system, conventional or otherwise, can substitute for good nutrition.

3. If nutrition is good to excellent, most of the common health challenges and stressors will not adversely affect rabbits because they'll be healthy enough to take them in stride.

4. Conversely, any rabbit herd, regardless of management strategy, will be plagued by disease, parasites, poor growth and poor reproductive performance if their diet is lacking.
5. Any given rabbit owner will be well rewarded for investing time and effort in finding and maintaining a high quality feeding program. That investment is the single biggest guarantee that the rabbit herd will perform up to expectations.

6. A purchased feed will go a long way towards ensuring a balanced diet, but any given purchased feed has gaps.

7. A variety of foods, in the generally correct proportions, is the best guarantee that a rabbit herd is getting the nutrients they need.

8. The question of whether the feeds are purchased or home-grown is actually a question of cost efficiency and personal preference, rather than an issue of rabbit health. Vitally rich vegetables grown in a home garden might have more nutrients per gram than conventional vegetables purchased in a store. However, even purchased vegetables offer nutrients which are not available in a standard dried pellet.

9. The complete diet as recommended on the HRS website (detailed in the Diet section) is the single most "complete" diet we've ever seen listed, anywhere, and in our opinion is an optimal combination of sound science, practical availability, user-friendly and rabbit-friendly. A rabbit owner would be hard pressed to improve on those recommendations.

Infectious Disease: Of all the reasons to not use colony management for rabbits, infectious disease is usually in the top three (the other two are parasitic infections and rabbit aggression, which we'll address in turn). In our experience, infectious disease can hit conventional rabbit operations just as often, and just as hard, as colony-based rabbit operations. Because so many infectious diseases are spread through the air, through touch or through water, the conventional cage design of commercial rabbitries offers very little protection against any of those modes of disease proliferation. In other words, rabbits are just as likely to get an infectious disease if they are in solitary cages, as they are in a colony environment.

That being said, infectious diseases can actually be easier to prevent and treat in a colony environment (with some caveats) than in a conventional rabbitry. This is because many immune-boosting supplements, whether herbal or synthetic, can be introduced via the water supply or feed supply to a colony with relatively little hassle. That same treatment would need to be given to every single individual rabbit in dozens or even hundreds of individual water bottles or feeders, or via an automatic watering system, for a conventional rabbitry. The one caveat to that statement is when vaccines must be injected into each animal (which can be the case with tularemia). In that instance, a colony may or may not be easier to manage, depending on how easy it is to capture individual rabbits within the pen. This is an issue of pen design, not indicative of the colony management concept in general.

Parasitic infections and control: another of the biggest typical arguments against colony management is the concern about parasites. The theory is that rabbits kept together in a group pen will concentrate any one rabbit's parasitic load such that all the animals become infested. Commercial rabbit cage design, by contrast, allows the rabbit manure to pass through the floor and thus not allow spread of parasites from one animal to another. While commercial cage design does minimize animal access to manures, those same cages present other health issues which can be just as devastating to overall rabbit health as any parasitic load might be. Furthermore, the statement that parasites "might" be shared between colony rabbits does not automatically mean that they will be. As with any herd species, the real issue is how often the herd is moved from one location to another for mobile groups, or how often the bedding is changed for stationary groups, and whether those moves/bedding changes are sufficiently frequent to break the parasite life cycle. If the bedding is changed (for permanent pens) or moved (for mobile pens) every few days, that single management strategy is sufficient to break the life cycle of any parasite which may infest a single rabbit and thus effectively minimize or eliminate any possible health impact. Furthermore, most animal species can easily accommodate a low level parasitic load without any negative impact, if their nutritional program is good.

Sore hocks: this health issue can sometimes be a minor problem for rabbits in conventional cages, or it can become a chronic, even life-threatening health problem. We have found that getting rabbits off wire flooring entirely, and putting them either on a deep-bedding system or a pasture environment, will clear up sore hocks usually within a few weeks.

Fur pulling: fur pulling is normal for a doe about to kindle, but it is not normal otherwise. When it occurs amongst either males or females who are not about to kindle, it is a behavioral response to stress. The stress may be a result of environmental conditions (such as chronic predator pressures) or social conditions (such as loneliness). We occasionally saw this situation when we kept all our rabbits in individual cages, but the problem went away entirely when we switched over to colony management.

Hygiene: this is sometimes listed as a reason to not keep rabbits in colony settings. Specifically, the statement usually goes that you can't keep rabbits clean in a colony environment. That is purely an issue of pen design and management, rather than whether rabbits are kept in colony groups. Any given cage or pen, whether occupied by single rabbits or groups of rabbits, can become a mess of mud, manure and urine if neglected long enough. That same cage or pen, with common-sense design and management, can be kept clean enough that the occupant(s) never experience hygiene-related health issues.

Aggression between rabbits: This is one of the big three reasons for not keeping rabbits in colony environments. I'll get into this in more detail shortly, but for the moment I'll summarize by saying that aggression prevention can be minimize or entirely avoided by understanding how rabbits interact, and which combinations are likely to produce fighting. We'll cover that in more detail in a separate section further on.

Reproductive management: one complaint about colony management is that it's hard to manage any given doe's reproductive performance. Specifically, many people are concerned about the timing for breeding, the provision of a safe nesting area, and the protection of newborn litters within a colony environment. We have already shared our practices along these lines, and I can say that those practices have proven to be easy enough that they don't create a compelling reason to stop using colony management. In fact, I have found that colony management actually makes consistent breeding easier. That's because I have sometimes gotten so busy that I missed signs that a doe was in heat and ready to be bred, and I thus missed that window of opportunity for another litter. Additionally, the steps that I currently take to separate out the doe, and create the nest box to protect the newborn litter, I would have done anyway. So there's no real increase in workload.

Social Behaviors which are encouraged by colony management: this has been one of the single biggest payoffs we've seen since switching over to colony management. Watching rabbits who each live in individual cages leads one to believe that rabbits naturally sit around like lumps all day. Watching those same rabbits in a colony environment, and they are transformed into living breathing interactive creatures. They will sleep in big furry rabbit heaps, each one draped over another. They'll groom each other, chase each other, play with each other, and snug up against each other. Furthermore, they'll explore the boundaries of their pen, they'll play with toys or pen furniture, and they'll stand up on their hind legs, periscope-style, to take the occasional look around. If that colony is managed on pasture or earth even part of the time, their behaviors expand again to include digging, browsing, nibbling, and exploring even more. Anyone who says that individually housed rabbits do all the same behaviors as rabbits in a group, must never have actually observed two such collections side by side.

Social behaviors which may be aggravated by colony management: All the wonders of social rabbits aside, there are some behaviors which can be instigated or made worse by colony management. The first is rabbit aggression and fighting. This is most often between two bucks which come into contact with each other, but it can also occur between two dominant does who are fighting for that top social position. As we have already detailed, both of these situations can be minimized or flat out eliminated by managing the adult populations in ways which preserve the social order, and minimize upheavals.

Breeding, Pregnancy and Kindling Issues
1. Rabbits require 14 hours of light to start and maintain a pregnancy. In the wild that isn't a problem, because they simply breed during the late spring/summer/early fall seasons. In captive commercial operations, interior lighting is usually already available simply for the sake of the personnel tending the rabbits, so this issue doesn't come up. yet day length becomes an issue for breeding rabbits when they are outside, on a property which is

north of approximately the 40th latitude. That's roughly the northern half of the country.

Locations north of this latitude have such short days in winter that reproduction grinds to a halt for half the year unless supplemental lighting is provided. If your breeding pairs or pens aren't producing during the winter months, check your light source (or start providing one). As a very general rule of thumb, if you don't have enough light to read by, for 14 hours/day, your rabbits don't have enough light to breed by.

2. If bucks are kept together in the presence of does, they will often fight viciously with each other and can cause fatal injuries.
3. A standard grouping is a single buck with 2-8 does. Any more does than that and the buck may not be able to keep them all serviced, or he'll show favoritism with one or two and ignore the rest. An ideal seems to be 3-4 does per buck.
4. If multiple groups are kept on the same property, separate them by more than a single fence line. Males will try to fight through the fence and may either injure themselves, tear up the fence, exhaust themselves trying to reach the other male, and/or ignore females in heat.
5. Underweight and overweight does can have problems with either ovulation and/or pregnancy maintenance. Keep the does at a healthy weight for best ovulation/conception changes.

6. Doe weight can be done most precisely via a calibrated scale, but it can also be done quickly via lumbar scoring. This is perhaps the more practical approach since many females can be evaluated very quickly without the hassle-stress of putting them on a scale.

7. If allowing a litter of young rabbits to grow out together, separate them into gender groups by the age of 2 months. Males and females are both capable of successfully breeding each other after the age of 2 months. The youngest mother we've ever had here was 3 months of age at the time she delivered.

8. If maintaining several breeding colonies, consider swapping the bucks back and forth a few times between colonies. That provides for more diversified breeding opportunities. Give the bucks animals time to adjust to each other prior to their first expected or scheduled breeding.

9. To control breeding times/pairings, one option is to keep the does in a group cage and then the buck in a solitary cage immediately next to the does. They can visit through the cage wall and experience most (but not all) of the benefits of true mixed housing. Then individual does can be removed from the group pen, introduced to the male in his pen, and removed after breeding.

10. If housing the buck separate from the doe(s), always bring the doe to the buck for breeding. If you bring the buck to the doe, the doe will attack the buck as an intruder to "her" space.

11. We have experimented with does kindling in both a group environment and in solitary cages. We prefer to house the does together for the most part, but then separate out the doe a few days prior to her expected due date (typically about 24 days prior to kindling) and then let her build a nest in a conventional cage and nest box setup. This ensures that we can check both the mother and the kits with the least fuss. It also ensures that we know which litter belongs to which mother.

12. Nest boxes:
    a. We prefer nest boxes with solid sides, an open top and a mesh bottom. This provides the best combination of retained heat, easy access and good drainage of urine away from the kits.

    b. The cardboard nest box liners from Bass Equipment will become sodden and heavy well before the kits are weaned. We started to replace them by about the 2nd week of life, but then stopped using them entirely and went with a mesh floor instead.

    c. The mesh on the bottom must be no more than ½" by 1" because kits can squeeze and wiggle through a 1" x 1" opening. Half-inch hardware cloth works well and is easy to replace. Standard ½" by 1" rabbit flooring material allows some bedding to fall through, lessening the amount of insulation between the kits and the outside air.

d. Standard ½" by 1" rabbit flooring also allows rodents to tunnel up from underneath the nest box, and they may nibble on the kits. We've had kits lose part or all of their feet due to rodent injury of this type. If rodents are a major problem, a mesh size of ¼" is usually sufficient, but 1/8" is better. Window screening is too flimsy. It can both tear, and be chewed through by rodents. Placing a solid barrier under the kits can work, but then the bedding must be replaced more often.

Sexing, Population Management and Herd Grouping
One of the topics which continues to come up for us with rabbit colonies, is group management. By that I mean how best to manage groups of rabbits to optimize their overall health, productivity, and social enrichment. Meanwhile we also want to optimize feed resources (whether purchased or home-grown), infrastructure resources (generally consisting of both the pens the rabbits occupy, along with the shelters which house the pens).

Last but not least, we also need to optimize our time and efforts that we spend on the rabbits, so that our rabbit operation meets our goals as efficiently as we can manage. Trying to combine all those elements together is roughly similar to juggling a dozen little balls all at the same time. Focus too long on one of them or even a few of them, and you'll end up dropping a few others. Yet focus a little on all of them and a balance point can usually be figured out. That is one definition of group management.

For rabbits in particular, there are a variety of ways to manage groups, but a few have risen to the top as being the most practical for us:

1. Sexed litters: when things are running smoothly, we sex our rabbit litters at the time of weaning, or roughly 1 month of age. I will say right now that we do not yet have a 100% success rate with that operation, either in terms of timing or accuracy. Thankfully, we get it right often enough that any "ooops" animals (i.e., one male mixed in with a lot of females, or vice versa) will usually become rather clear in the next few weeks. We try very hard to have complete separation of the two sexes by the age of two months. When we have gotten lazy about that separation, we've had kits born to does as young as three and a half months of age. Definitely not a good situation, since the doe is still growing herself. When we do stay on target time-wise, we then manage those two sexed groups together from that point forward, until either slaughter, sale or breeding.

2. Bachelor groups: our young males from sexed litters go into what we call bachelor groups. This is a group of anywhere from 2-8 young males, all of very nearly the same age, housed and managed together. We usually end up either selling or butchering most of our young males before the age of six months, so we rarely ever

have long-term housing issues or socialization issues to contend with.

3. Maiden groups: this is our term for groups of young females from sexed litters. Like with the bachelor groups, we maintain maiden groups of similar age, and we do sell or butcher some of them. The percentage that we sell is usually determined by three very different variables:
   a. how much demand we have from customers at any given time

   b. how much space we have available for them to move into, and

   c. our current rabbit herd size, and whether we want to expand or shrink that herd

   It's difficult to summarize what happens to any given maiden group because we've tried a number of different management techniques. We'll touch on this point repeatedly in the rest of this section. But for now I'll move on to our other grouping types.

4. Established breeding groups: this is generally a single buck with anywhere from 2-8 mature does, who are in various reproductive stages. This group tends to be a lot more dynamic over time, because we also generally pull a near-term pregnant doe out of the group setting and give her a solitary cage within which to kindle. She then stays in that pen for the next four weeks, until weaning. Given that setup, any given doe won't actually

physically be in a group setting for any length of time, since she'll by cycling in and out as she delivers.

However, we try to keep each doe relatively close to her group for that entire time, for instance by putting her cage immediately alongside the group pen. That way she can still go nose-to-nose with group members, without having to mingle directly. We like to think that cuts down on the disruption when we add her back into the group later. Maybe it does, maybe it doesn't. We do still get fights sometimes when we re-introduce a doe, particularly if she was a dominant doe and then someone else took her place during her absence. To date we haven't really worked out a foolproof formula for this portion of rabbit politics, but we continue to work on it.

5. Solitary bucks: as we have experimented with a variety of social groupings, one of the issues we have is how best to balance out the genetics amongst our various bloodlines. While we're not breeding for any particular color or type per se, we don't generally want brother/sister or parent/offspring combinations. So at some point, we have to swap bucks between groups, or bring in a new buck after selling/butchering another.

These transition points can often be a source of upheaval for the social order of any given group, so we've started using something of a "halfway house" approach to ease the transition. For instance, if we have two breeding groups, one with Buck A and the other with Buck B, we may decide that we want to swap those two bucks. Instead of simply taking each out of his group and then putting him immediately into the other group, we put them both into solitary cages for about two weeks. That timing interval is rather arbitrary, but it seems to be just about right to allow the does in each group to settle into the idea that the previous buck isn't coming back.

Once they reach that realization, then the appearance of a new buck seems to be less disruptive overall. We also try to house each solitary buck near his future group, so that they can start to get to know each other through the pen fencing. We have even occasionally put a buck cage right into a doe pen, so that they can all mingle around his cage without anyone actually getting too friendly too fast. That approach is helpful when we have a really dominant doe in the group who might otherwise attack him if he simply suddenly appeared out of nowhere.

6. Bonded Pairs: occasionally we will have two animals who become so strongly bonded that they would have severe stress if separated. We have occasionally seen that with sisters, mother/daughters, and with a buck/doe combination. The sisters pairing is usually not a major issue since they'll eventually be the nucleus of a new breeding group anyway, and thus will be together for the duration regardless. The mother/daughter pairing is a little tougher, because usually any given breeding group is single-generation. We have occasionally formed new breeding groups based on mother-daughter pairs and a single buck, if we

had some compelling reason to do so, for instance if the mother was an excellent breeder and the daughter seemed likely to carry on that tradition. For the buck/doe bonding, we have occasionally allowed two strongly bonded individuals to simply stay together, typically by letting the buck stay with the group in which the doe was already a well-placed member of the hierarchy.

Improving Herd Genetics Over Multiple Generations

The toughest situation we've faced so far in terms of group management, was when we really started running out of room and had to cut back the size of the herd. It was tempting to take the best performers from each group, and form a new group of "over-achievers". However, putting several dominant females into a single pen in the hopes that they'll just "work it out" has rarely ever given us good results. Usually one of them will stay dominant, and the others will continue to fight it out. That took its toll on reproductive success, such that we don't think we ended up creating our hoped-for group of overachievers. Given that, the next time we had to whittle down the herd size, we took it down one entire group at a time. That didn't look as good on paper, since we kept the low performers in the group as well as the high performers. However, there was much less wear and tear involved.

Given that history, I will do things a little differently if and when the time comes again that we have to whittle things down. In our previous downsizing, the realization that we needed to downsize came on rather suddenly, and we didn't have a lot of time to prepare. Having gone through that exercise twice now, I better understand the symptoms of a rabbit herd which has grown too big, and I hope/expect to be able to see that point coming a little sooner in the future. What I plan to do next time, if there is a next time, is to ID the top performers in each group, then breed them at least one more time, all at the same time. Then when those kits are weaned, we can group together all the young females from all the top performers in one super-group. By bringing them together at such an early age, we'll eliminate most of the fighting risk, while still keeping "the best of the best" from our previous groups. Then we can either sell or butcher the previous groups, secure in the knowledge that we've saved the best genetics for the next generation. We can take it a step further at the next breeding, by seeing which of the next generation actually delivers on their "best of the best" expectations. We keep those, and cull the rest. That would be an ideal way to steadily improve herd genetics over time.

Section 3: Colony Rabbit Enclosure Designs, Functions and Management

Introduction

I was debating with myself how best to present the information in this section. Having scribbled down dozens (if not hundreds) of notes and ideas and layouts over the last 15 years, only to modify and adapt and reject and refine those ideas, I wasn't sure how much detail I really wanted to get into here. I also didn't want to get into any sort of "this is the ONLY right way to do such-and-such". I have so often thought we had a design really figured out, only to learn some new way to improve it later. So what I've tried to do is distill down all the things we've tried over the years, into observations and results and comments about what worked (or didn't work) for our particular situation. Instead of a narrative format, this section will be a lot of lists, featuring the keynotes for a lot of different concepts. We hope that by presenting this information in this manner, you can pick and choose the elements which might work for you.

Pen Materials

1. We have used a variety of pen materials, varying from various fencing and cage materials, to jumbo dog carriers, to plywood stalls to conventional wooden hutches.

2. We have used both individual and group pens which rested on the ground, along with pens which were suspended above ground.

3. We have worked with rabbits at liberty within a yard enclosure.

4. A note about wire fencing descriptions: most fencing and cage materials of any kind, can be classified in one of two different categories.
    a. Welded wire is any wire mesh or fencing where the individual wires are welded at the joints. Hardware cloth and rabbit cage materials are both welded wire mesh. These types of fencing are most vulnerable to stresses at the joint, which can break the weld. Once that happens, that particular joint cannot be repaired even though the wires may remain in the same general orientation. Over time, that broken joint will work open either due to flexing of other areas of the mesh, and/or animal pressures against that particular opening.

    b. Woven wire is any wire mesh or fencing where the individual wires are fastened to each other by wrapping another piece of wire around the joint. The joint is then held together via friction. This type of fencing holds up to stresses better, but only to a point. If enough force is applied to that joint, or to the wires immediately around the joint, the wires will simply start to slide past each other and thus start to deform the mesh. That opening can then start to open wider and wider until it can become 2-3 times its original size. Under extreme pressures, several parallel wires may all slide past each other and open up very large gaps. At worst, the wires will simply experience metal fatigue at that point and break. Woven wire mesh can be repaired if the wires are still intact, by repositioning them back into alignment and then using same-diameter wire hand-wrapped around the new joint. That new joint will never be as strong as the original, and that piece of fencing will be prone to new damage from that point forward.

5. Our cage designs have varied from conventional ½" x 1" rabbit wire individual cages, up to group pens housing 20 rabbits at a time.

6. Our current favorite design is for pens which rest on the ground, stay put for a year or so, and measure about 4' wide by anywhere from 10' to 20' long. We clean out each group pen periodically, depending on the population density within, and depending on the feeding methods we were using. For instance, our deep bedding system feeds hay on the ground so that whatever isn't eaten will form a deep bed of warm, fertile mat which can then be blended into the garden soil underneath. I can't call this material compost, because it doesn't heat up, and I can't call it soil, because it's much more compacted than garden soil. But it is chock-full of fertility and does the garden bed a world of good. We move these pens from one garden bed to the next each year, such that the rabbits build up each planting bed intensively during the year. We keep the group pen covered such that both the rabbits and their interior surfaces are shielded from rain, snow and wind.

We take comfort from the fact that the pens all have bottoms such that the rabbits can't dig out, yet their urine soaks into the ground where those nutrients will be used later for garden vegetables.

7. Wire fencing material findings:
   a. Rabbit cage wire (i.e., 1" x 2" welded wire mesh) is the most expensive of the various potential pen materials to buy, but it is just about the ideal to work with for all rabbits except newborn kits. It is strong, long-lasting, easy to cut to size, rigid enough to stand up to years' worth of field use, yet light enough to move easily. We used to buy it by the roll from Bass Equipment, but now we just buy it from our local feed store. That saves us the cost of shipping.

8. Several of our tried-and-vetoed pen designs:
   a. 2" x 2" woven wire field pens: too floppy such that they needed external supports. Difficult to move. The opening size allowed kits to escape. The flooring would seem to allow access to grass underneath, but it was so matted down that the rabbits had a hard time reaching it. They would generally give up after a day (or less).

   b. Concrete mesh (4" openings) forming an arch over which rabbit wire was fastened, with rabbit cage wire (1/2" x 1" mesh) for flooring: much too heavy to move, and the flooring virtually guaranteed that any grass would be matted down. The ends were also hard to close off due to the arched shape. We used hardware cloth which didn't last very long.

   c. We tried lightweight materials such as chicken wire and hardware cloth in a variety of settings. Those materials did OK if they were suspended above the ground such that they didn't stay wet very long, and if they were not subject to frequent moves. Those materials did not stand up very well to frequent movement, exposure to high-moisture situations such as low along a pen wall, or direct contact with the ground. In those settings, they either tore or rusted out within a year.

   d. So-called 2"x4" welded wire field fence makes for generally good large paddock fencing, except it must be reinforced along the base because the rabbits will push against it while trying to get that just-out-of-reach blade of grass. That opening size will also permit passage of young rabbits through the fence without any trouble at all. We are currently re-fencing our garden area, which currently houses our rabbits, with a combination of 2"x4" welded wire field fence, combined with 1"x2" rabbit wire along the lowest foot of fencing. We expect we'll have to replace both forms of fencing within 10 years, but it should be relatively rabbit escape proof until replacement.

9. We have experimented with many rabbit-tractor designs, and we weren't satisfied with any of them

enough to keep using them. Some of the problems we encountered:

a. The lighter and easier to move they were, the faster they broke down in use. They also became more prone to rabbit escapes as wall-to-wall or wall-to-floor joints worked open or simply broke apart.

b. The stouter they were, the harder they were to move. The more they weighed, the less inclined we were to move them. That led to dramatically less efficient usage of pasture than what we'd planned.

c. We had problems both with the fencing materials we used, and with the methods for joining them. For instance, the J-clips will work open over time if a joint flex during movement. So-called zip ties, made of clear, white, black or colored nylon, work well as quick connectors but most of them are not UV-stable, so they break down with a year of exposure to direct sunlight. They are also favorite chew-toys for the rabbits, who will chew on them whenever they can reach them.

d. Light-gauge wire such as hardware cloth and particularly chicken wire, will separate or simply break at join points.

e. J-clips proved most durable, but we needed a lot of them (one clip approximately every 3"-4") to ensure that they didn't work open.

10. We got tentatively positive results when we laid down wire on the ground and then allowed grass to grow up through the mesh. The rabbits got excellent browse without being able to dig out. However, our very uneven ground made it difficult to match up the cage sides with the bottom cage mesh, such that we often had gaps at the bottom. Those gaps turned into rabbit escape points all too frequently. We never did find a way to ensure that the tractor pen bottoms mated up perfectly flat with the wire on the ground. The best we could do was level the ground in advance, try to keep the grass growing as evenly as possible, and position the tractor to minimize the inevitable gaps between wall and flooring. That meant a fair amount of fiddling around with the tractor after moving it to the next desired location.

11. We have taken a break from the rabbit tractor plans for the moment, given the tractor issues listed above. However, we are currently working on implementing something of a hybrid design which we believe will address most if not all these issues. Our plan is to house the rabbits in relatively immobile group pens like we already do, and then surround those pens with planted grass areas lined with heavy-gauge galvanized wire such as 2"x2" woven field wire. By planting several different "rabbit pastures", we can rotate the rabbits through each growing area, while also ensuring that pen fencing is securely connected to the ground wire. We also know that we can direct rabbits from one area to the next by opening or closing gates to and from the

different areas. In this sense, the rabbit paddocks would simply be miniature rotational grazing areas commonly used for other grazing species, with a series of aisles connecting a centralized housing area to a series of satellite yards.

12. We know now that we can count on standard rabbit cage wire and 2"x2" woven field fence will last roughly 10 years before rusting out. We have also figured out that we can use 24" wide rabbit cage wire, and when the bottom 2" rusts out, we can simply trim off the rusted portions and attach new flooring, thus dramatically extending the life of the rest of the cage wire. We have gone as short as 18" before the rabbits stop some of their vertical movements like upwards stretches. At that point, we switch that wire over to serve other purposes, such as temporary pens for young growing rabbits, or non-rabbit uses. One thing we've experimented with is to use old 18" rabbit wire as a barrier in our garden against the cats getting into the planting beds. Given that our climate is quite wet, and our soils stay moist far longer than in some areas of the country, rabbit wire fencing (or any fencing for that matter) might hold up quite a bit longer in dryer climates. Anything which allows the metal to dry off quickly after getting wet will lengthen its functional life.

13. We no longer use wood in any capacity for the rabbits, given our early experiences with wood hutches. We have also largely retired the use of nylon zip ties except for emergency fence repairs.

Pen Design
Now that we're reviewed a variety of pen materials, how best to design the pen itself? That again will depend on a lot of different variables, such that no one pen design is a "fits-all" for each possible situation. I should also note here that we've built a lot of different pens, both portable and stationary, and we have yet to come up with a design that we feel fits all our needs, all the time. If we haven't developed a single design that always works for us, I can't realistically claim to be able to recommend a single design for others. Instead, I'd like to make some observations and recommendations based on the various designs that we've tried.

1. Square footage: Rabbit colony pens don't need to be extremely big. If you consider that most commercial cages for rabbits are only 18" x 18" square, or even 24" x 24" per rabbit, that's only 2.25 – 4 square feet per rabbit. So, the area of a sheet of plywood, which is 8' x 4', would be enough for eight rabbits. Granted, many people want to give rabbits a lot more room than what commercial cages usually afford. Even with that goal, a pen does not need to be very big to allow for plenty of exercise and socialization activities. We have found the 4 square feet per rabbit to be a sufficient space for both social activities, along with giving each rabbit the chance to get some personal space. The one instance where this is NOT enough space is when introducing new rabbits into an established group, because the newcomer can't get away. Since we try to minimize those instances, this usually isn't a problem. However, on the rare occasion

when we do need to introduce a new animal to an established group, we do so in a much bigger area, let everyone get used to each other, then put them back into a smaller pen.

2. Shape (Square vs rectangular vs round versus oblong): This might seem a rather silly detail to look at, but it has major implications for cost and practicality. One item thing to consider is that any given pen must not only give the rabbits enough room, but it must also give you the chance to reach into the pen, either to make repairs, handle the rabbits, or place/remove things like feeders, waterers, hay racks, etc. Let's consider the above example of a 4' x 8' pen, housing 6 to 8 rabbits. With those dimensions, you can reach the mid-point of the pen from the sides relatively easily, because the mid-point is only 2' away from either of the long sides.

However, let's now consider enclosing that same square footage in a square shape. On paper that would seem to be a good idea, since it would be slightly less fencing materials to enclose the same space. Each side would be just over five and a half feet long. However, that means the middle of the pen is almost 3' from the sides. A rabbit or other object right in the middle of the pen, would be out of reach from the sides. You'd have to climb into the pen to reach it. If the pen is tall enough to walk into, that's no problem. However, if the pen is only rabbit-high, you'd have to lean over from the sides. That gets very uncomfortable very quickly.

Another issue is corners. For some reason, predators look first to corners to enter a pen or yard. Perhaps they have learned over time that corners are a traditionally weak location where gaps might be found, or the fence materials more prone to movement and thus to squeezing through. One poultry owner acquaintance of ours has unbelievable coyote pressures on his chicken tractors when they're in the field. Those pens are typically far from the house and not in a high traffic area, so coyotes can "work" the pens for relatively long periods of time without anyone noticing. He used to have rectangular field pens but found that the coyotes would consistently break into the pens at the corners. So, he switched to round pens.

Shortly after making that switch, he had occasion to watch a coyote trying to gain entrance to the round pen, by watching via binoculars. He said the coyote circled around and around the pen and couldn't seem to figure out where to start working the pen materials to gain entrance. After 30 minutes of this, the coyote gave up and trotted off to look for an easier meal. He kept that pen design from that point forward, and as far as I know he's still using that design for that very reason.

A third issue is moving the pens. To touch on the concept of "rabbit tractors", many people will want their pens to be portable, while others won't ever move the pens. That is a purely personal decision based on a variety of details. Generally, though, if you plan to move a pen on a regular basis, a square or rectangular shape will be easier to move simply because it can more easily be made rigid. One way to gain the best of both worlds, would be to have a small central square or rectangular hutch which the rabbits can be put into for moving day, then have a roll of fencing to surround them. That small hutch can be moved from spot to spot, and then the roll of fencing set up around the hutch to allow for a much larger grazing area. Then prior to each move, lock up the rabbits in the hutch, roll up the fencing, and you're ready to move. This approach would require the use of fenceposts in high-predator-pressure areas, so that predators didn't simply walk over the top of the fencing. Relying on fenceposts could be a minor inconvenience or a major hassle depending on any given owner's priorities, so once again that would have to be a decision made based on those priorities.

3. Height: When I talk about pen height, I'm referring to two completely different issues. Namely, the amount of vertical spaces the rabbits have, versus the question of whether the pen is intended for you to walk into comfortably.

   In terms of rabbit comfort, pens and cages of any size intended for rabbits generally need to be at least 18" tall, but 24" is better. Giant breeds like the Flemish, need at least 30" tall. That height allows for the rabbits to sit upright with ears erect, without banging their ears into the pen roof. So, commercial cages for Californians or New Zealand breeds are often at least 18" tall; field pens are similarly usually 24" to 36" tall. Also keep in mind that rabbits can easily hop out of a cage or pen which has sides 24" tall, so a pen of that height normally has a roof on it of some kind. A 36" high pen is getting a little tall for rabbits to hop out of, but cats and dogs can readily get into a 36" high pen without a roof over it.

The question of whether to build the pen tall enough for people to walk into, is going to be determined in large part by other variables. For instance, if a pen is being built inside a room, garage, barn or other existing building, then portability is no longer even an issue, and the pen can be as tall as the building which contains it. There are a lot of advantages to having a pen tall enough to simply walk into. Tasks like handling the rabbits, cleaning up, repairing/replacing/refilling feeders and waterers, etc. is all easier when you can simply walk up to each item. However, very generally speaking, a person-tall pen is a stationary pen. If you want that pen to move, a much shorter height will be substantially easier to move for reasons we discussed above. The one exception to that is the example where only a small central hutch is moved, and a roll of fencing is used to provide grazing area. If that roll of fencing is 24" or 36" or even 48" tall and has a roof on it, you won't be walking around inside. If the fencing is 6' tall, or doesn't have a roof, walking around inside is no problem.

4. Weight: this is really only a concern when the pen is intended to be portable. After all, the pen must be rigid enough to hold together during each move, yet be light

enough to move. The method of moving the pen will determine how rigid and how heavy it can or should be. Many rabbit tractors are designed to be movable by one person. Commercial versions of portable pens are often made of lightweight materials such as chicken wire or rabbit wire, stretched over and fastened to a rigid frame of either metal or wood. Some commercial and homemade pens are much heavier, and designed to be moved via vehicle or tractor of some kind. In that instance they might still be movable by one person, but that one person must drive the vehicle. In that instance, any animals within the pen should be contained within a hutch or other fixed carrier, since the person can't drive and shepherd the rabbits simultaneously.

A middle-of-the-road approach is to use something like a hand truck or dolly to move a pen along the ground for short distances. This as the benefit of getting at least part of the pen off the ground, maybe even an inch or so, and thus make it easier to slide or drag along. Rabbits can be allowed to follow along in that instance since the person moving the pen can simultaneously watch the rabbits to make sure no one gets run over. Another option is to put the pen on wheels, skids or runners, such that it's easier to drag from one location to another. Many traditional hutches on stands can be retrofitted with either wheels or skids so that they can be made portable. Even relatively heavy traditional wooden hutches can be moved short distances in this manner by two people.

5. Roof or no roof? Rabbits will need shelter from sun, wind and rain in some form. That shelter can take many different forms, ranging from a simple tarp over a small area of the pen, to a pen completely enclosed within a conventional stick-frame room or building. The decision is often based on what infrastructure is already available to the rabbit owner. Many rooms, spaces or buildings have been adapted for use as rabbit pens. The main concerns here are that the rabbits have somewhere to escape inclement weather (which for rabbits can include hot, dry, sunny weather), and that predators can't get into the pen. For predators, consider not only terrestrial predators such as cats and dogs, but also avian predators such as hawks and owls. Also consider that the roofing does not need to be the same type over the entire pen. A large, well-fenced yard area could feature a tarp over one corner, and have the rest of the yard open, with lower objects like tables, chairs, benches, shrubs, etc. serving as cover against avian predators.

The only other issue to consider with roofing is that water cannot be allowed to flood the pen area. Damp or muddy ground will chill rabbits and make them extremely prone to illness very quickly, i.e. within hours of initial exposure. So make sure their shelter area is also high and dry such that rainwater or snow can't accumulate underneath. Also, if using lightweight materials like a tarp, make sure the tarp slopes away from the yard such that water runs off out of the yard. Finally, be aware that a tarp which is only supported along the edges, will sag in the middle and can collapse under the weight of either snow or ponded rainwater. A tentpole or ridgepole can keep this from happening.

Pros and Cons of Portable versus Stationary Pens

In the previous section I talked about a lot of different pen designs, and observations we made while using them. In this section I'd like to summarize the pros and cons of two general types of pens, namely portable versus stationary. I will say up front that I don't intend to recommend one over the other. Rather, I'd like to present their strengths and weaknesses so that folks can decide for themselves which pen would work in their particular circumstances.

Portable Pen Pros:

1. Can move the rabbits to fresh ground, either for grazing/browsing purposes, to prevent mud, provide better hygiene, clear patches of ground, or simply for enrichment purposes.

2. The rabbits can work any one area more completely, then move on to the next area in a systematic way, rather than turning them into a large area and having some areas over-used and other areas under-used.

3. Different groups of rabbits can be kept far away from each other, or near each other, as indicated by whatever goals or challenges may exist.

4. If property usage plans change, the pens can be moved to more suitable locations.

5. If an operation moves to a new property, the pens can go with them.

6. Manure and urine can be allowed to soak into the ground, thus enriching the soils within the pen.

7. Given the above, rabbits can be used to "terraform" new garden spots by simultaneously weeding it, digging up the soil, fertilizing the soil and then moving on before compacting the soils too severely. They can also be used to provide a very natural, well-balanced lawn fertilizing program, provided they are not kept on any one patch of lawn so long that they graze it or trample it to death.

8. Predators are often thrown off by pens which are moved every few days. This can reduce or sometimes even eliminate predation pressures. One note that the longer the pen stays in one spot, the more likely predators will have a chance to study it and find some weakness to exploit.

Portable Pen Cons:
1. Any pen left too long on a single piece of ground will result in denuded earth, trampled and/or compacted soils, possibly muddy soils and muddy rabbits.

2. The rabbit owner must do the planning and prep work to ensure that the rabbit pen always has a new spot to move into, for each portable pen in use. If using only one pen that isn't a huge issue, but when managing several pens that can become quite a management burden.

3. Constantly changing pen locations can make it tough to carefully balance the feed ration. Either the feed ration needs to be modified each day to allow for the rabbits' consumption of grass with differing nutrient values, or the pasture (or lawn) must be managed more proactively to ensure that the rabbits are kept on grass at the same stage of growth. Either way, it's a lot more work than simply feeding out hay and pellets.

4. Owner illness or injury, or unforeseen duties such as working late, kids' after-school programs or even visitors from out of town can throw off the moving schedule, which in turn may also throw off the rabbits' nutritional program.

5. The rabbits will be using a lot more of the property over the passage of time, and that area may need to be kept free of other activities even when they're not using it. That can result in a lot of "real estate" that doesn't seem to be in use, but which must still be paid for (either in terms of rental, or in terms of property taxes, mortgages, etc.). That situation can be manipulated to an extent by allowing other livestock to use those areas when the rabbits aren't there, but that in turn will complicate management of the area in general.

6. Any predator control measures, such as electrified fencing, roofs, blinds, etc., will have to move with the

pens. So each time the pen is moved, there might be several different items (hutch, fence, shelter, etc.) which each need to be moved. That may complicate and/or slow down the move operation, or turn it from a one-person job into a two-person job. Conversely, if the pen is an all-in-one structure, that entire structure will have to be designed and built to be movable.

7. Any provided feed, and all the water, will need to be taken to the new pen location regardless of how far away it is from where those materials are stored. This might not be an issue if using a single pen in a suburban yard. This might be a huge issue if maintaining multiple pens in a pasture situation, when the pens eventually reach the far end of the pasture.

8. Escaped rabbits will generally stay close to familiar ground. However, if they have covered a wide range of pen locations over a short period of time, they may tend to range further since all of that previously covered ground is "familiar". This will also depend on how much cover there is in the immediate area, such that rabbits may not tend to wander very far in a wide open pasture environment. If moving the pens through a brush-filled or landscaped environment, that might result in escaped rabbits ending up quite far from where they started.

Stationary Pen Pros:
1. Permanent flooring can be built or used to ensure the rabbits don't tunnel out of the pen. Depending on what this flooring is (concrete, tile, brick), it might make cleanup extremely easy.

2. If a permanent pen is made from standard rabbit wire supplies (with standard rabbit flooring of ½" x 1" mesh wire) the entire pen can be suspended over worm bins. The rabbits plus the bins will result in two benefits from a single area of the property. This can be a crucial advantage in extremely small areas which must maximize productivity.

3. A permanent pen, hutch, yard, room or building can be heavily invested in once to make it as suitable for the rabbits as possible, then not need much repair or maintenance from that point forward.

4. If neighborhood predators are a problem, strong defensive measures can be built once and then simply monitored and maintained, rather than continually moved and re-inspected.

5. Similarly, if neighborhood dogs are a problem, blinds, soundproofing or other protections against dog-induced stress can be installed once and then simply maintained over time.

6. Chores can be made very efficient by setting up feed storage immediately next to the permanent pens, and having a water source nearby. This minimizes the amount of time spent doing daily maintenance.

7. Permanent pens can be carefully designed to accommodate known weather risks, such as dependable heat/humidity issues in summer or wind-driven rains and snows in winter.

Stationary Pen Cons:
1. Manure and urine buildup, either in dirt flooring or on solid flooring, must be addressed on a periodic basis. The longer that task is put off, the bigger job it becomes. If ignored, that situation becomes a health hazard for the rabbits, and possibly a pollution source or public health issue.

2. The herd size is constrained by the size of the permanent pen. If the herd grows dramatically, new quarters must be constructed. If the herd size drops, the pen (and possibly the building) might go under-utilized.

3. Predators and vermin (such as mice and rats) will learn every square inch of a stationary pen or building's infrastructure, and will repeatedly try to find ways into the pen(s). This can become an ongoing management headache.

4. If property usage patterns change, that stationary pen might someday be in the way even when it was originally located in the perfect spot. Similarly, a permanent pen might be ideal right up until the property needs to be sold or rented to new occupants, at which point the pen becomes an eyesore or unwanted hassle. Rabbit owners using rental properties in particular may need to carefully consider how best to build a permanent pen, with an eye towards how to quickly dismantle it and return that space to its original condition or purpose.

One Design Option: Seasonal Pens

The section above, detailing all the pros and cons of both portable and movable pens, may seem a little overwhelming simply because there are no slam-dunk answers. If you're trying to decide between mobile versus stationary and they seem equally appealing, consider a hybrid approach. Specifically, a seasonal system where a stationary pen is used part of the year, and portable pens are used another part of the year.

This approach works quite well if your location lends itself to portable pens for some portion of the year, for instance rabbit tractors on pasture during the growing season, yet winter conditions require the stouter protection of a stationary building. We have that situation here, with sometimes harsh late fall, winter and early spring storms of wind, rain, snow and sometimes freezing rain. Not only does the weather require stout shelter, but our water table comes way up and we sometimes have standing water in pasture areas. Rabbits in portable shelters would be severely stressed for the entire winter season under those conditions, and their productivity would suffer accordingly. Yet our late spring/summer/early fall conditions are perfect for the rabbits to be out on pasture in lightweight shelters which can move to new grass on a frequent basis.

So our management plan has slowly evolved to a seasonal system. We pull all the animals off pasture around the first of October, as our fall rainy season really kicks in, and keep them in well protected stationary housing during the winter. Then as the weather improves in spring, we turn them back out usually around the end of April or early May. This approach gives us the best of both worlds.

Security and Rabbit Escape Issues
Regardless of what pen materials, designs or layouts you use, inevitably you'll have rabbits escape. It is perhaps a law of livestock husbandry that somehow, someday, someone is going to get out. So instead of leaving you to debate how best to respond after the little miscreants have already escaped, I wanted to share some information up front for how to minimize those escapes, and then how to deal with them when they do occur.

Our observations on common escape methods and situations:

1. Adult rabbits can squeeze through 4" by 4" openings.

2. Weaned rabbits can squeeze through 2" by 4" openings, and even some 2" x 2" openings such as chain link fencing.

3. Kits can go through standard 1" x 2" rabbit cage material, and their limbs can go through ¼" mesh if bedding is scarce, wet, and/or if the kits start digging.

4. One good strategy for minimizing escape issues is to have two barriers between any given rabbit and freedom. Often a rabbit will test the nearest barrier, particularly if bored. However, they won't test the same size barrier in a larger containment area quite so fast, because they'll be too busy exploring. That's an opportunity to catch them.

5. Rabbits will typically go under a fence by tunneling, or squeeze through small openings between fence panels (such as in corners or near gates/doors.) Reinforce fencing edges in corners and around gates/doors, such that the fence panel edges are rigid enough that they won't bulge open if pushed against. An extra panel of fencing which overlaps a corner or gap between gate and fence, will prevent escape at those points. Ending a fence against a solid rigid barrier such as a vertical fence post can also prevent rabbits from squeezing through.

6. Rabbits won't generally jump over fencing any taller than 3', but they can and do stand on their hind legs and stretch up with their front paws to see how high a vertical surface goes. If they have the opportunity to thoroughly explore an area, and they're bored and/or brave, they can learn to jump higher than 3'. If they can climb a smaller item such as a bench, low table or even an overturned pot alongside the fence, they can then use

that as a step to get over the fence. One simple way to ensure they don't test low fences, is to make sure objects aren't left laying alongside the fence that they could hop up onto, and then climb the fence from there.

7. Rabbits won't generally climb a fence by putting feet through each individual opening and using those openings as a ladder.

8. Most of our rabbit escapes have come from rabbits squeezing under a fence. They can do this even when the fence goes right to the ground. They succeed with this whenever the fencing material is not tight/rigid enough along the bottom, that they can push through. Even a 1" gap at the bottom can be pushed open to several inches if there is even a little slack in the fence. Also, they will exploit minor dips in the ground level to find a spot to squeeze under. They will also commonly burrow under a fence.

9. One way to defend against this is to extend the fence down into the ground by 6". Another method is to have the fence bend into the yard at ground level such that they can't get under it and can't dig through it. That fencing would have to extend several feet into the yard, AND be stapled or weighted down so that the rabbits can't simply wriggle underneath it to reach the actual fence line. A third alternative is to simply lay fencing throughout the pen so that they can't dig anywhere, and so there are no gaps between ground level and vertical fence. A fourth option is to let grass grow through

permanent fencing laying on the ground, then slide bottomless pens along that fencing to new grazing areas.

When Rabbits Escape: What to Do Next

The escaped rabbit will willingly explore a relatively large and new-to-them area. We have had rabbits in pens in a 30' x 60' fenced garden area explore the entire yard within a few hours of their escape, even though they may never have been outside the pen before. They will often go from one hiding spot to the next, spending relatively little time exposed and visible. However, some rabbits seem oblivious to the fact that they are out in the open and thus visible. That seems to vary on a rabbit-to-rabbit basis.

Some of our rabbits have been friendly enough that when they escaped, they simply hopped up to us upon our arrival in the yard. This has happened even with rabbits who were rarely if ever handled in the past. Don't count on that as a recovery technique, but don't be surprised if it happens.

The easiest way to catch a loose rabbit is to use the paddock approach. If they are already in a contained area, then simply start dividing the area into smaller and smaller paddocks until they are in a small enough space that you can reach them without a big stressful chase. For instance, in our 30' by 60' garden area, we keep a roll of 30' long, 24" high fencing rolled up along the 60' side. That fencing is out of the way for normal use, but it's ready to be unrolled to the other side of the garden if a rabbit happens to escape. By doing so, we've cut down the catch pen by half. We have additional rolls of 24" high fencing staged in other areas, ready to be deployed in similar fashion wherever they are needed. We just keep moving each roll to create smaller and smaller pens until the rabbit is in a space roughly 4' x 8'. If the rabbit is tame and has been gently, calmly corralled, that pen size is usually sufficient to simply reach over the side and pick up the rabbit. If the rabbit is nervous or easily excitable, we have a catch cage which is basically just a big rabbit-wire cage of roughly 3' x 2' in size, with a bigger than usual door. The door can be latched open by flipping it over the top of the cage. We put that cage in a corner of whatever paddock, usually with some rabbit treats in it. When the rabbit hops into the cage to check out the treats, we casually walk up and close the door.

Another way to catch a loose rabbit is with the funnel approach. In this method, several very long rolls of 24" fencing are needed. The fencing ends at a catch cage like the one described above, such that the rolls of fencing "funnel" the rabbit into the cage. When the rabbit escapes, the cage and rolls are then deployed in such a way that the rabbit is not directly approached at first. The cage is set up some distance away (about 2/3 the length of the rolls of fencing), and then each roll of fencing is deployed off to each side of the rabbit. This forms the walls of the funnel. Then the rabbit is approached from the "mouth" of the funnel, at which point it will typically move directly away, and thus further down the funnel. These first few steps towards the funnel mouth (i.e., the catch cage) will give a person time to bring the far ends of the two rolls of fencing together, such that the rabbit is now completely contained. At that point the person is in the catch pen with the rabbit, and only needs to casually, slowly move towards the rabbit to encourage it towards the catch cage. Once in the catch cage, close the door.

All capture actions need to be done quietly and calmly to keep the rabbit from bolting. Any quick, loud or sudden motions on the person's part can flush the rabbit into a blind run in any direction. If that happens despite everyone's best efforts, stop and take some time to cool down (at least 20 minutes). That will allow the rabbit to calm down too, so that a subsequent capture attempt will go more smoothly.

If the rabbit has escaped and disappeared, stay calm. It is probably nearby and watching you from under some hidey-spot. Long grass, shadows, an overturned flower pot, between hay bales, and blackberry brambles have all served as hidey-spots here. Give the situation some time and see if the rabbit materializes over the next few hours. Of all our escapes over the years, none of them have ever gone more than 100' away from wherever they started.

If the rabbit has escaped next to a busy road, DO NOT ENDANGER YOURSELF OR OTHERS by going out into the road during a chase. You can easily get distracted and get hit by a car yourself, or cause a car to swerve and create an accident in the effort to avoid you. Rabbits are important but they aren't THAT important.

We do NOT recommend the use of any kind of net to capture escaped rabbits. We have found that the netting will spook the rabbits and cause them to panic. That in turn will cause them to start kicking and struggling against the net. Either the net will tear, or the rabbit's legs will get further entangled in the netting, thus panicking the rabbit even more, or they will kick so hard against strong netting that they'll snap their spines. It can happen in the blink of an eye. We strongly encourage folks to use either the paddock or funnel recapture approaches rather than use netting of any kind.

If you do use netting to capture a rabbit and it becomes spooked, you MIGHT be able to calm the rabbit by turning it on its back and keeping it there. This is unproven and only to be used as a last resort. The theory is that by putting the rabbit in that position, you'll put it into a hypnotic state which will then allow it to calm down such that the netting can be safely removed. While we have hypnotized rabbits this way for various medical and testing procedures (such as gender testing), we've never used it under such stressful circumstances as a recapture. As such, we can't guarantee that it will work. Proceed at your own risk, but consider it if you've run out of other options.

Section 4: Colony Rabbit Environmental Issues

This section is intended to be something of a catch-all for issues which didn't come up under the previous three chapters. As you'll see in the topics, many of these issues span rabbit colony diet, health and housing. Instead of trying to lump them under one of those three categories, I decided to let them stand on their own. Hopefully that will make them easier to find for folks just looking for this-or-that piece of information. Given that these topics are a compilation of observations on our part, I'm going to move away from the narrative writing and just list our observations, for folks to work through as they see fit. As usual, I'll provide more resources at the bottom of each section.

Predator Issues
1. Most people keeping rabbits don't have direct predator or predation problems, in the sense that they don't have predators killing and carrying off rabbits from the herd. However, predators can and do cause indirect pressures on colony rabbit herds by their mere presence. A coyote or fox circling the pen incessantly may not cause bodily injury, but they can and do cause heart attacks, litter absorption, stress-induced illness and other indirect responses.

2. Predator issues can include wildlife, such as coyotes, raccoons and raptors. It can also include neighborhood or household pets such as cats and dogs.

3. Predators do not need direct contact with the rabbits in order to negatively affect rabbit health, stress levels, health issues and/or reproductive success.

4. Predator protection will ideally go well beyond simply keeping predators physically out of reach of the rabbits. Predators should also be kept out of visual range so that they can't stare at rabbits for long stretches of time.

5. If a household dog or neighboring dog barks constantly, that can also (but not always) impact rabbit health. Finding creative ways to either keep the dog from constantly barking, or give the rabbits an "auditory break" from the sound, will give the rabbits a chance to "come down off red alert".

6. Rabbits in a field pen situation should be given plenty of sheltering places to hide, so that they can get out of sight from either terrestrial or avian predators.

7. Wild rabbits seem to prefer fairly brushy country, with 40% ground cover. This mix of open ground and cover gives the rabbits a lot of choices for grazing/browsing in the open, while never being far from cover.

8. This amount of cover can be simulated in open pens by providing things like low tables, boxes, barrels, shade cloth or other visual screens and solid objects. Just make sure they are safe to chew on.

9. Cats may pose the single greatest complication to rabbit owners:

a. They can be extremely persistent about getting into rabbit areas. While dogs can usually be fenced out of rabbit enclosures, cats are much more nimble and can go through very small openings just like rabbits can. However, they will easily jump over fences that rabbits would never jump.

b. Add the complication that many homeowners welcome cats as a viable form of rodent control. While the cats will readily hunt rodents in the rabbit area, they will also readily hunt baby rabbits.

c. Finally, stray cats and true feral cats are so common, that almost no portion of the country is without at least some risk of cat predation.
d. If cat predation seems to be a likely or known issue, the single best way to protect against them is to house expectant doe rabbits, and their litters, in a separate area or building. This will help ensure that the presence of cats doesn't cause abortion or savaging, and will protect young rabbits against outright predation.

10. Stray dogs and coyotes, and to a lesser extent foxes, may not be as common a risk as cats but they can require more efforts at control if they are present.
    a. They will put much greater pressure on fencing to get into rabbit pens.

    b. Their mere presence can be enough to make does abort litters, and/or savage young kits.

c. A dog, fox or coyote that can pace the pen fencing within sight of the rabbits, can stress the rabbits so badly that the rabbits either bolt into an immovable object and hurt/kill themselves, or keep their stress levels so high that the rabbits suffer heart attacks and die without the coyote, dog or fox ever getting into the pen itself.

d. Two ideal defenses against known canine presence, are strong fences which the canines cannot go over, under or through, and a visual barrier so that the canines can't "stare down" the rabbits. A hotwire around the outside perimeter of the fence can be a powerful deterrent to canines against pushing through or digging under. Fences should be at least 4' high to discourage climbing and/or jumping over. A pair of hot wires, one approx. 6"-12" from the ground, and a second wire at or near the top of the fence, will deter all but the most determined canine predator. Be aware, however, that a starving canine will risk much, even bodily pain and harm, to get a meal. If hungry canines and/or packs of roving canines (wild or domestic) are a known issue in the area, bringing the rabbits into a building may be the only way to ensure their safety.

11. Raptors are typically only a problem for rabbits if the rabbits are kept in large outdoor pens (approximately 20' x 20' or larger). The birds need a certain width in order to dive down from flight or from some nearby

perch. They also need a certain "launch distance" to be able to take off again with a heavy rabbit in their talons. If the rabbits use a large yard for either exercise or grazing, fill the yard with lots of objects which the rabbits can hide under for protection against avian predators.

Pest Issues
Rodents can cause almost as many problems for rabbits as predators; in some cases more.
1. Rodents can and do learn to eat rabbit foods, and most rabbits are timid enough that they will move aside if a rat approaches the feeder.

2. Rodents can spread disease amongst rabbits, particularly a disease called tularemia (aka rabbit fever). This disease can wipe out an entire rabbit herd in a matter of days.

3. Heavy rodent pressures can stress out rabbits such that they don't breed, don't conceive, don't carry to term and/or savage their litters.

4. Rodents will chew on wooden structures and nest in deep-bedding systems, right under the rabbits if allowed to do so.

The single most effective way to control rodents is to rodent-proof the rabbit pens so that the rodents cannot gain entry. While we can't predict from here where your rodent-sized gaps are, one helpful way to review your operation is to look for all the places where a young weanling rabbit could squeeze out of your pens or enclosures. If a young rabbit can squeeze out, rodents can squeeze in.

If rodent pressures are already high because they're in the area for some other reason, deploying additional control measures (trapping, poison or even hunter cats) might be necessary (but see our comments on cats and rabbits).

Tularemia has been the single biggest herd health issue we've ever experienced here in our 15 years of keeping rabbits. And that disease was almost certainly introduced to our operation via rodents. It can be spread by physical contact, so any materials which rats have access to can become a vehicle for transmitting the disease to rabbits. Some suggestions specifically to control tularemia:

1. If rodents are anywhere near the rabbit areas, be sure to wash hands before/after handling the rabbits, any rabbit equipment/feeders/fencing, or hay/bedding and other materials.

2. Wash feeding/watering equipment if bringing that equipment into the rabbit area after possible exposure to rodents.

3. Do not let rodents congregate in hay or bedding storage areas, because their feces/urine can spread the disease on the hay and bedding.

4. A vaccine against tularemia is available for rabbits; veterinarians can administer the vaccine or may be able to dispense it to an owner for vaccination at home. Local and state laws will impact whether vets must be

the ones to administer the vaccine. If tularemia is a known or suspected issue in your area, discuss the possibility of a vaccination program with your vet.

5. If a number of rabbits suddenly develop fever and then die, or die very quickly without overt signs of illness, suspect tularemia as a strong possibility. The disease can move very quickly through a herd, but it requires several weeks to incubate prior to the appearance of disease symptoms. If a group of rabbits suddenly die in a group (within a few days of each other), and then another herd on the same property follow that pattern within a few weeks of the first, tularemia is a very strong possibility.

Pasture Management
We addressed the nutritional aspect of this topic in the chapter on Diet and Nutrition. In this section, we'll talk about managing the plant communities of a grazing area, from the point of view of promoting strong plant growth, harvesting that growth in a controlled way, and rotating grazing areas to ensure long-term plant stand health and vitality.

When we use the word "pasture" we are talking about any area which features plant growth of any kind, regardless of the area, the plants being grown or the percentage of the plants which the rabbits are likely to browse. So the word "pasture" in this context could include lawns, gardens, typical household landscaping (even houseplants if a rabbit colony is housed indoors), a woodland paddock, or a conventional pasture in use by other livestock.

Rabbits put three different kinds of pressure on a pasture:
- Trampling pressure, by their mere presence on the grasses

- Grazing pressure, by eating the grasses, forbs and other plants in any given area

- Digging pressure, whenever the rabbits try to dig either depressions or tunnels in the soil.

Any plans to graze or browse rabbits must anticipate the above three pressures, and have a plan to either minimize or manage those pressures so that the grazing area can be maintained over time.

Trampling Pressure:
1. This category of pasture damage will depend on the time of year, the moisture content of the ground, the growth stage of the plants, how hard the rabbits graze down the plants, how many rabbits are in the area and how long the rabbits are there.

2. There's no way to objectively predict how much trampling pressure any given area can take, given all the variables listed above. Some common sense, though, can give us some general predictions. The more rabbits are in an area, the smaller that area, and the longer that the rabbits are there, the more severe the trampling damage will be. Moisture content becomes an issue in two situations: if the area is very wet, the rabbits can trample living plants down into a muddy mess much

faster than if the ground is slightly to moderately wet. Similarly, if the area is extremely dry, the rabbits can trample dried plants down much faster than if the plants had normal levels of moisture.

3. There will always be some trampling losses when rabbits (or any animals) are turned into a growing area. The biggest two questions are:
   a. How much trampling is acceptable? This will depend entirely upon the owner's preference, and can vary widely. If the rabbit owner wants those rabbits to eat everything in any given area with minimal waste or leftovers, and thus minimize trampling losses, one of the single best strategies to do that is known as managed grazing. Unfortunately, managed grazing, and all its variations, is much too big a topic to go into here. Those interested in learning more can read up on various managed grazing methods at http://extension.psu.edu/plants/crops/news/2015/04/the-benefits-of-managed-grazing-systems. We'll also provide some additional resources at the end of this section.

   b. Is that damage temporary or permanent? Most forms of rotational grazing aim to steadily improve pasture quality over time. We'll get more into this topic in the section on grazing pressure. Just know that most models for rotational pasture management assume a certain amount of trampling damage. For now, know that most plants (whether grasses, forbs or shrubbery) can sustain some trampling without permanent injury. In fact, some plants are stimulated to grow faster by such damage. One very general rule of thumb that we use here, is that most plants can easily recover from damage to 25% of their overall mass through either trampling and grazing. While this can be hard to measure objectively, it can be estimated visually by inspecting the growing area at least once a day. If 25% of the plants have broken stems, broken or nibbled growing tips, matted stalks or other signs of damage, it's time to pull the rabbits off the area and let it recover.

4. Experimentation over time is the best way to determine how much trampling pressure any given plant community can sustain. Start slowly, and start with very low animal densities for very short periods of time. See if any given growing area is visibly damaged after the rabbits have come and gone. Also see how long that area takes to recover with new plant growth. Then increase from there. Your own observations will start to tell you where the balance point is.

5. Also keep in mind that this balance point is always a moving target; it will depend upon the season, the temperatures, the precipitation, etc.

Grazing Pressure:
1. This is damage caused by nibbling on the plants rather than standing on them. From the plant's point of view,

the damage is very similar because the nibbled or trampled materials will probably not regrow, and thus the plant must send out new growing tips. The difference, however, is that trampling is purely wasted plant material. Grazing provides a direct benefit to the rabbit in the form of nutritional intake. So there's always a tradeoff between protecting the plant while giving the rabbit(s) adequate nutrition.

2. As with trampling pressures, that balance point can be difficult to estimate in advance. When assessing a pasture's nutritional content for feeding purposes, the first question to ask is whether the entire plant can be taken, or only some portion of it. In other words, does the pasture need to survive beyond that single feeding session? If the answer is yes, the pasture needs to survive for future feedings, then only a fraction of the pasture can be taken at any given time. How much can be taken? Again, that topic goes well beyond the scope of this writing. However, some guidelines do exist:
   a. Most grasses cannot be grazed lower than 3" without seriously damaging the plant's ability to regrow. So any estimates of pasture yield should include only the length of grass blades longer than 3". In other words, if a pasture is 6" tall, the only half of that grass depth can be grazed before the rabbits need to be pulled off and moved elsewhere.

   b. Forbs (wide-leaved forage plants such as plantain, lettuces, most leafy greens, dandelions and other wild plants) have the same tolerance for pruning that we mentioned prior, namely about 25%. That means that up to 25% of the plant can be removed (either through trampling or grazing) and the plant can easily recover. Heavier grazing, however, will start to seriously weaken the plant.

c. Legumes have a similar tolerance rate of about 25% as other plants.

d. One important note: it can be helpful to review grazing guidelines for other species, but keep in mind that other species graze in different ways. For instance, cows, sheep and horses all graze at different depths when given the choice. However, any animal, including rabbits, will graze the entire thing all the way to the ground if hungry enough. For the vast majority of forage plants, that graze-to-the-ground will kill the plant outright. So while monitoring rabbit grazing, keep in mind that they may simply kill a certain percentage of the plants in any given area through grazing if allowed to do so. The longer they stay on a piece of ground, the more plants they'll take all the way down to soil level, and the more replanting will be needed afterwards.

Digging Pressure:
1. Rabbits will dig for a variety of reasons, including the creation of shelter, the need to create a nest, and sheer boredom. Some soils seem to stimulate rabbit digging, while other soils seem to discourage it (for instance, very

sandy soil doesn't seem very interesting to rabbits for digging purposes).

2. The only way to absolutely prevent rabbits from digging is to keep them on concrete. They will attempt to dig everywhere else, including on wire mesh.

3. Rabbits can be discouraged from digging to a certain extent by providing them with pre-existing shelters and nest boxes. When on soil, they will still attempt to dig even when wire mesh is on the ground. In that instance, they'll dig up small patches here and there and leave the mesh exposed. If the priority is not to prevent the digging, but rather to prevent an escape, then wire mesh on the ground will suffice, and allow grasses to keep growing through the mesh.

4. If high-value plants or landscaping are growing in any given area, such that digging cannot be tolerated (for instance, under prize shrubbery, or in working orchards or vineyards where roots should not be disturbed), fence the rabbits out away from the individual plants. Or fence the rabbits out of that area entirely.

5. Even when indoors rabbits will attempt to dig. Their most common victims are potted plants, carpeting and upholstery. While carpeting and upholstery damage is usually an issue only for pet rabbits, indoor rabbit pens may very well have a number of potted plants in the room. To protect these plants, cover the exposed soil with hardware cloth or other stout wire mesh.

Alternately, a wooden lid of some kind can be used, which allows for watering but is closed otherwise.

As we indicated above, grazing science is a huge topic, which goes well beyond our scope here. For more information on various grazing management approaches, check out these resources:

A 38 page PDF from the University of Missouri's Cooperative Extension Service: http://extension.missouri.edu/ozark/documents/2014_Grazing_School/IntrotoManagementIntensiveGrazing.pdf

A shorter, 8-page introductory PDF, this time from the University of Vermont: https://www.uvm.edu/~susagctr/resources/IntensiveGrazing.pdf

Fodder Growing Systems

One very common concern amongst urban rabbit owners is that they want to give their rabbits access to fresh greens, but either have extremely limited access to the outdoors, or no access at all. In this instance, one creative solution is to grow grasses in long, low rectangular growing trays and then cover the trays with hardware cloth. The grasses can easily grow through the hardware cloth, and the rabbits can eat the tops of the grasses without trampling or digging into the potted roots. If rabbit owners wanted a continuous supply of such greens, a number of these trays would be needed to rotate in and out of the rabbit pen. For instance, a 9" x 24" rectangular pot with 4" of grass above the hardware cloth could be brought into the rabbit pen in the morning, and then grazed all day. It could be removed to a sunny window to regrow while another tray was put into the pen the next day. Again, depending on environmental conditions, the regrowth period would vary. But it could be as short as 10 -15 days if the grasses had sufficient heat, light and liquid nutrients. Even giving rabbits such greens once every 2-3 days would provide a wonderful nutritional boost, while reducing the needed number of trays by half or two-thirds. And such trays would virtually eliminate all the trampling, grazing and digging risks we've discussed in this chapter.

If the planting trays are too much work, another alternative is to grow fresh greens without soil of any kind. Essentially, the greens would be overgrown sprouts. The entire plant – green leaves together with the roots – could be fed to the rabbits. This is perhaps the most labor-intensive option of all, but it is a wonderful option for those with extremely limited space. Once again, this topic goes well beyond the scope of this writing, but you can get more information on growing sprouts for livestock feeding purposes below:

Commercial fodder production systems:
https://www.farmtek.com/farm/supplies/ExternalPageView?pageKey=EXTERNAL_PAGE_3017
http://www.foddersystems.com/

DIY fodder production systems:
https://www.peakprosperity.com/wsidblog/80359/diy-home-fodder-system
http://www.theprairiehomestead.com/2016/02/fodder-system.html

Weather Issues

Rabbits can tolerate a wide range of environmental conditions. Yet they have a few specific requirements which must be met. Thankfully, those needs are usually fairly easy to provide in some way. Here's a hodge-podge of observations about helping rabbits combat whatever extreme weather situations might be in your neighborhood from time to time:

1. Rabbits are extremely cold-hardy. They are very comfortable in temperatures all the way down to the teens (in Fahrenheit).

2. They are not so comfortable in hot weather. Temperatures above 80F will cause rabbits to reduce their activities, put up their ears for maximum heat loss, and stretch out in their cage or pen to maximize the body's ability to radiate heat. Temperatures above 85F will cause rabbits to start panting. Sustained temperatures above 90F can kill rabbits due to overheating.

3. Hot, dry conditions are slightly easier for rabbits to tolerate, since by panting they lose a small amount of body heat. Hot, humid conditions are more difficult because panting is not effective in that situation. In either situation, several items can help rabbits cool off:
   a. A big fan moving a lot of air will help the rabbits cool off, by using their ears as radiators. Moving as much air as possible past their ears will maximize that potential heat loss. However, air speeds much beyond 5mph (in other words, much beyond a large house fan) could be irritating to them and cause them

to lower their ears, such that any such advantage is lost.

b. Ice packs to lay on and/or frozen water in water bowls to lick, can both help lower body temperatures.

c. A misting system in the environment, but NOT on the rabbits themselves, is a good way to cool off air temps in hot and dry conditions. However, the higher the humidity, the less effective it will be.

d. A cool, dry dirt, brick, tile or concrete floor will allow the rabbits to stretch out and radiate heat directly into the ground. Note that the surface must be cool to the touch in order to be effective. Sun-baked surfaces will only make the situation worse.

e. Indoor rabbits greatly appreciate air conditioning on hot muggy days. They'll sit right in front of the air conditioner and let the cold air sweep over their ears until they are comfortable again.

f. Rabbits can survive short spells of hot weather, i.e. hot afternoons followed by cool evenings, nights and mornings. If temperatures are going to stay above 70F continuously, plan to provide them with one or more of the above temperature control measures.

4. Rabbits absolutely, positively need protection from getting wet. The more wet a rabbit gets, the more likely

it is to get sick and die within a matter of days. Wet feet and belly fur earned on dewy grass will easily dry off on a nice summer afternoon. However, getting soaked during even a summer rainstorm can kill them. Getting wet during the fall, winter or spring can kill them even faster. Do not let rabbits get soaking wet, and do everything possible to not let them even get damp. For outdoor rabbits, always give them a place to get out of the weather.

5. Wind is not normally a problem for rabbits, unless it is bitterly cold. During extremely cold, windy weather, they'll curl up in a ball with their ears tightly held against their body. A windbreak of any kind can make a huge difference to their comfort levels under such conditions. A burrow or other container can provide a very comfortable place to ride out a storm.

6. The only other time wind is a problem for adult rabbits, is in circumstances of wind-driven rain. A very small amount of rain, with enough wind, can soak a rabbit very quickly. That can kill rabbits faster than just about any other weather condition other than extreme heat. For outdoor rabbits, this can usually easily be prevented by ensuring that they either have containers to withdraw into during stormy weather, or by ensuring that any pen includes enough roofing or overhang to protect against wind-driven rain or snow.

7. While we've been talking about adult rabbits so far in this chapter, we should point out that newborn kits are

completely unable to regulate their own temperatures. Their nest box bedding mix of dry grasses and Mom's fur are normally sufficient to provide all their insulation needs. However, that mix must be kept absolutely dry in order to maintain those temperatures. If the bedding gets wet for any reason, the kits will die very quickly. If a newborn kit gets out of the nest for any reason during anything less than 50F, it will die very quickly. If kits are born "on the wire" or out of the nest box for any reason, warm them up to 100F as quickly as possible. One of the best ways to do that is to put them in a closed Ziploc bag, with plenty of air in the bag, and then float the bag in a bowl, sink or bucket of 100F water for a few minutes. Sometimes apparently chilled and dead kits can be revived this way.

Section 5: Colony Rabbit Business and Regulatory Issues

Cost Effectiveness

This topic may or may not be an issue for any given rabbit owner, thanks to a number of variables. For folks with a small herd of rabbits, say one buck, a few does and their offspring, cost effectiveness may not even register on the list of things to pay attention to. However, if/when the herd size starts increasing, or a herd owner starts selling rabbit products, suddenly this topic starts to come into view. Similarly, this topic may not have much priority until/unless budget constraints start to demand attention. At that point, suddenly cost effectiveness can be the first thing you think of in the morning and the last thought in your head at night.

Cost effectiveness can cover a variety of topics within the realm of keeping rabbits:

1. How much are you paying for feed?

2. How well is the feed being converted into meat, fiber, reproduction or overall health?

3. Would it truly be cheaper to grow your own feed (and we're counting time, money and energy here, for ALL the associated activities of planting/cultivating/harvesting and storing home-grown feed).

4. Is any of the feed (purchased or homegrown) being wasted? That could include (but not be limited to)

things like spoilage during storage, spillage at feeding time, waste once the feed has been fed out, and poor uptake/metabolism after it's been eaten).

5. How much time/money/effort has gone into building infrastructure, and how long does it last? Similarly, how much maintenance and repair does it require during its useful service life?

6. How expensive are the feeders and waterers to acquire, and how durable are they? Do they lend themselves to fast and easy chores, or do they require extra effort?

7. Are there elements in the environment which are creating problems for the herd, resulting in higher costs or effort? For instance, a neighbor's barking dog or a high-traffic area could result in stressed out rabbits, with resulting lower reproductive success, poor weight gain, and poor overall health. Could these environmental issues be addressed in some way to reduce their impact?

Any/all of the above issues are almost negligible for a small herd with only a few rabbits. However, once we start to really depend on those rabbits for meat, fiber, manure and/or income, suddenly these issues become more important.

While as an author I can't peer into any given operation and make suggestions, here are some cost effectiveness issues we've faced in the past. Perhaps they will serve as either good suggestions for things to look for, or at least as warnings for what types of issues can negatively impact a rabbit operation.

1. When we first got started, we bought fully assembled cages for our rabbits. When we realized that they were extremely simple to build, we bought the cage materials from a home improvement center and started building our own. Then we realized that the home improvement center charged a lot more for those materials than other retailers, either local or online. We then started buying the cage materials by the roll from the cage manufacturer (Bass Equipment, Inc), but we had to pay shipping for each heavy roll. Finally, we realized that we could order the same materials through our local feed store, and pay for only the roll itself. The feed store still marked up the roll a little bit to cover their own shipping costs, but their costs were substantially lower than by going through the home improvement center, or paying full shipping to get it from the manufacturer. All told, our cage costs now are roughly 1/5 the cost that they were when we started, simply by looking for more cost-effective sources and methods.

2. Daily watering may not seem like a huge issue, until you have 100 rabbits in July. Then it becomes a major part of your chores cycle. We started with the fairly standard half-gallon plastic water bottles, one for each individual cage. As we moved over to group cages and then to colony pens, we kept the water bottles going with each new cage or pen design. We learned that a single water bottle would need daily refilling for a group of six or more rabbits. Two water bottles for that same size

group would both be half-full the next day, but each rabbit had much easier access to it because the dominant rabbits wouldn't guard both simultaneously.

When we had multiple group cages going in the field, we experimented with a more involved watering system composed of a 5-gallon bucket which drained into something like a drip irrigation system. We purchased just the metal watering nipples from Bass Equipment (as of this writing, those nipples are still available for such purposes) and used black irrigation line to run a string of multiple nipples out from the five-gallon bucket. That system looked good on paper but we never got it to work very well.

We originally wanted to use clear tubing, but we knew that clear tubing would soon become clogged with algae. The black tubing didn't have that problem but then we couldn't see where there might be a blockage. We had to individually test every single nipple every single day to make sure it was still delivering water to the pen. If we had a blockage somewhere, we learned it was simpler to just cut more tubing and replace the entire section, rather than try to find the blockage. The barb fittings we used to join the different sections would sometimes leak, particularly if the cut end of the line wasn't perfectly perpendicular to the tubing wall.

We also learned that rabbits are very entertained by trying to pull the tubing into the pen, and then shredding it into a million little pieces. We never did figure out if they ingested some of that plastic while shredding it; we assume they probably ingested a little. The fact that they also drained the five-gallon bucket in the process was apparently part of the entertainment value.

Finally, after finding ways to defend against all those issues, we learned that a single hour of freezing weather will freeze that entire spaghetti bowl full of tubing, and it takes 3 hours to thaw. In the end, after a whole lot of time, money and energy invested in trying to get the automatic watering system to be truly automatic, we went back to the half-gallon water bottles.

3. When we got started with rabbits, we bought the standard rabbit pellet diet from the feed store. At the time, we had the option of buying a small 10# bag, or a much larger 50# bag. We started with the small bag, since we didn't really know how much feed we'd be going through at any given time. Once we had a feel for that volume, we started to look at the 50# bag because it was considerably cheaper per pound than the 10# bag. At first we were concerned about spoilage, but we were able to virtually eliminate the possibility of spoilage by putting the dry pellets into a clean metal trash-can with a close-fitting lid.

   Later as our livestock population expanded (both in number of species and in number of animals per species), there came a time when we were buying both the rabbit pellets, which are mostly alfalfa, alongside straight alfalfa pellets for the dairy animals. During that time we were also expanding both our gardening efforts and our hay feeding. When we first got rabbits, it would not have made much sense to start gardening or buying hay just for them. However, since we had started doing those activities for other reasons, it made very good financial sense to start 'sharing' the harvest with the rabbits. Then at some point along the line, it suddenly occurred to us that we didn't need to buy the dedicated rabbit feed anymore. The straight alfalfa pellets were quite a bit cheaper, and we had all the other ingredients we needed (hay, veggies, fruits, pasture, etc.) to provide for a complete diet for the rabbits.

So now our feed costs per rabbit have gone down quite a bit, and the burden of producing that feed here is spread amongst not only the rabbits but the other livestock which use the alfalfa, the hay and/or the veggies. While our workload went up overall during that time, the portion of that workload created by the rabbits was almost negligible compared to providing feed for the other livestock. So our net cost effectiveness for the rabbits went down with those changes.

The single most important thing to do on a regular basis (four times a year is a good baseline), is to look at the major money, time and effort "costs" for your rabbit operation, whatever size it may be, and see if there might be ways to reduce those costs. It's often a very good idea to at least compare/contrast different possibilities on paper before putting them into practice. Also, don't be afraid to experiment with different options to see how well they work out. Just keep those experiments small until they prove themselves. The several examples above are each good examples of how different options may either be advantageous, or not, as a rabbit operation changes over time.

Regulations

I hesitated to include a section on regulations, because for the vast majority of rabbit owners it won't be an issue. However, it is becoming a bigger issue for more and more individuals over time, so I decided that a short discussion was warranted. There are two general categories where regulations might kick in: animal welfare, and sale of rabbit products. We'll consider them each in turn.

Rabbit Ownership, Population Density, Welfare and Sanitation Issues

As of this writing (March 2017), I am not aware of widespread or specific regulations which bar the ownership of rabbits anywhere. They are small, quiet, clean animals which can easily be kept as pets even in apartment-type settings. Most livestock regulations are concerned with the size of the animal, the potential for either noise or smell issues which could bother neighbors, and/or disease issues which could impact nearby water quality and/or public health. And for most rabbit owners with only a few rabbits, that's as far as it goes.

However, once again the size of the operation starts to matter. If you've got a breeding trio in a small pen in your basement or garage or spare bedroom or back yard, hardly anyone is going to object. On the other hand, if/when those numbers start to climb, more and more people will notice. Depending on their observations, concerns and overall opinions about animals, that may or may not become a problem.

We've learned a few things along the way with livestock in general, and with rabbits in particular, which are of note in this context:

1. People have vastly different ideas about what constitutes "good care", "good housing" and humane treatment in general. Those opinions are frequently NOT based on fact, but rather on what people expect to see, and what experiences they've had in their own lives. For instance, someone who has no experience with rabbits might think that sub-freezing temperatures are a problem for outdoor rabbits. Or if they had pet rabbits as a kid, they might think that a rabbit in a cage is cruel. Or if they had rabbits in cages, they might think that rabbits in a pen is irresponsible. Whatever. So Rule #1 with people: don't assume that everyone will look on your operation and approve of it. They won't.

2. If/when people don't like what they see, they may feel compelled to tell you you're doing it wrong. That can come from either the next-door neighbor, or someone walking by or driving by your property, or even a visitor to your property such as the mail delivery person, a parcel delivery person, a guest, friends, relatives, etc. Everyone has an opinion and many people believe their opinions need to be shared with the world. Rule #2 then

is that people will often tell you what they think, even if they have no business doing so.

3. Sometimes people will be so concerned about what they see, and their opinions will be so strong, that they will complain about you to various local, state or federal agencies. Perhaps they think they're doing you a favor, or "saving" your animals from living conditions that they find objectionable. Some of those agencies will be compelled by law to respond with a site visit, just to check out the allegations. The rights and obligations that you have as owner of the animals and/or owner/tenant of the property will vary by location. This is where Rule #3 kicks in: regardless of what you THINK the rules should be, it is in your best interest to find out what the rules actually are. What you think they should be often has nothing to do with what they actually are.

4. While every individual has the option to follow or not follow any particular rule, it is at the very least a good idea to know what they are. So Rule #4 is: know your local applicable laws for animal welfare, animal housing, and/or animal care in general, and for rabbits in particular. This includes details such as how many rabbits are allowed per household, what types of housing or shelter they must have, what types of feed and water must be available, etc.

Our Visit From Animal Control in 2011

Perhaps it would be instructional to run through a situation we faced here a number of years ago, which may have been instigated because our rabbit colonies were visible from the busy road which borders our property. For the first 12 or so years that we were in operation as a livestock operation, we had very little concerns about any of this. We were doing basically what we knew we needed to do, namely providing clean feed and water, in sufficient quantities, to our animals. We kept healthy animals in clean yards with sufficient shelters, and made reasonable efforts to ensure they couldn't get out of their pens. In those rare instances when they did escape, we made darn sure they never left the property. And that was pretty much that. But then the day came when an animal control officer came to our farm gate, while we weren't home, and left a note for us that they needed to schedule an appointment with us to investigate charges of animal neglect. Excuse me?!? Apparently someone had filed a formal complaint about our operation, and that complaint required an inspection.

My first response was total disbelief. Someone was alleging animal neglect? Why? With which animal(s)?? We didn't have answers to any of those questions. My first call was to a friend and livestock mentor of mine who lives in a nearby county, and who is occasionally called to investigate cases of animal neglect in her county. We talked for a long time. She had been here numerous times and had seen how we keep our animals, and was able to reassure me that we probably didn't have anything to worry about. She explained that many times, a "good Samaritan" will file a report because they are concerned about what MIGHT be going on, not because of what they KNOW is going on. My friend offered to come down and walk the property with me and look at our operation through the lens of someone who is being consulted on an alleged animal neglect case. I accepted her offer. She came and spent the whole day with us, walking the property and explaining to us what criteria she used to evaluate people's animal operations. For the most part we were fine. We did have some very minor instances were some animal maintenance tasks hadn't been kept up to date, for instance the goats needed to have their hooves trimmed, the horse fence needed some repair and the cows' water tank could stand to be cleaned out. But that was it. She suggested that we call the officer back, schedule the appointment a few days' out in time, then take care of those items in the meantime. She also suggested that we be as cooperative with the animal control person as possible, since they were just trying to determine if anything was amiss. They probably weren't going to come onto the property and hassle us, contrary to a lot of Hollywood storylines to the contrary. So we followed her advice, scheduled the inspection for a few days out, then took care of the few items on the list.

The day of the inspection, I was really nervous, but things went pretty much the way she had predicted. The officer was very polite, and explained that they had gotten a complaint dated right before they had left the note on the gate. The officer also explained that he had looked at as much of the property as he could from the front gate, and hadn't seen any indications of problems, but he needed to see the rest of the property while we were at home. So we walked the entire property again, and he took notes of various things. He asked us a few questions about this and that, but found nothing of major concern. We asked him who had made the complaint and why, since we hadn't had any visitors to the property in some time. Since we don't have sidewalks in our rural area, but we do have a busy road out front. We couldn't figure out how someone would have seen anything amiss, driving by at 45mph. The officer explained that it might have been a neighbor who (ironically) was concerned specifically because he or she couldn't see our operation. Or it could have been a motorist driving by at 45mph who didn't approve of our operation simply because it didn't fit with his/her mental image of how animals should be managed. Of all our livestock groups, the only two groups which were readily visible from the road at that time were our goat herd, and our rabbit colonies.

He asked me how long we'd been at this location, and I said 12 years. He commended me that we'd gone so long without a single complaint, because usually someone will make some complaint about a high-visibility operation within just a few years of it becoming noticeable. I asked how we could keep this from happening again, since it had taken his time and my time and nothing had been wrong. He suggested that the single best thing we could do, was move our livestock yards away from the road, and/or put up landscaping or shrubbery along the road, so that people couldn't glance over as they drove by to see that we have livestock of any kind. In other words, out of sight, out of mind. He concluded the visit by saying that our operation was just fine. He would file his report to that effect, and that was the end of the inspection.

The lessons learned from that event?

1. It doesn't actually matter how well we keep our livestock; someone somewhere will disapprove. So while we are working hard to keep them according to our own standards, we should also invest a certain amount of effort in ensuring that our operation is either invisible to passers-by, and/or as attractive as possible to minimize the risk of complaints.

2. In our instance, the public servants who came out to do the inspection were very easy to work with, very reasonable in terms of their evaluation of our operation, and very cooperative in terms of helping us to understand how to minimize our risk of future complaints.

3. Knowing what our legal obligations are, is more important than "keeping up with the Joneses". We may not be able to make all the observers happy all the time, but as long as we are supplying the living conditions required by law, no one can successfully file complaints about us and make them stick. We might have to waste an afternoon going through inspections to prove that point, but if we're operating within the law, that's the end of the conversation.

If the moment comes when a paid public servant is at your door requesting or demanding to see the premises, your options get real short real fast. Many times, such a visit is not random; a complaint was made to a government agency about your operation and those public servants are required by law to check into it. Your options include:

1. raise a stink and yell and scream and refuse them entry, which may or may not work very well. Depending on location, they may or may not need a warrant to check your property, and refusing them entry could make matters worse for you.

2. On the other end of the spectrum you can let them in and give them full access to the property and not say a word. While that has the illusion of "cooperation", that actually is not very constructive either. Most public employees haven't yet made up their minds about what they're going to find. If you stand there like a statue and refuse to answer their questions, you're not making their jobs any easier. That refusal to answer questions will be reflected in their report about your operation.

3. Another option is to give them free access to the property and not even accompany them. This also is not generally recommended, because then you have no idea what they did or did not check. They may have missed some part of your operation which is an important part of proving that you're providing competent care. For instance, if you happened to pick up all the water bowls that morning with the intent to wash them and replace them later that day, that's a commendable management practice. However, if they notice that you don't have water bowls and you're not there to answer questions, they could accurately write down that you don't have water available to the animals. So you could make things worse for yourself by not being willing to talk to the inspectors.

As of this writing, none of the above options are recommended by the various public agencies that we've worked with over the years on such complaints (either complaints we've made, or complaints which have been made about us).

The current "best practices" procedure would be:
1. Allow the public servants entry, and cooperate with their inspection of the premises. Be as polite as you can manage to be, even though it will feel like an invasion of privacy.

2. While they are there, get their name, their badge number or employee ID number, the agency that they represent, and get it all on paper. If they have a warrant (some locations require that, and some don't), get the details of that warrant – the date it was issued, the agency or jurisdiction which issued it, any reference number, and a copy of it if possible.

3. Get as much information as you can about the complaint which was filed against you. Keep in mind that most agencies are not allowed to say who made the complaint, but most of them will tell you what the complaint was about (i.e., noise, smell, animal welfare concerns, etc.). The agency itself will often indicate the source of concern; for instance, a visit from Public Health won't be checking on animal welfare, but rather on potential disease issues. And Animal Control is often tasked with checking on animal welfare/cruelty. Keep

in mind that in rural areas, the Sheriff's department might have to wear multiple hats and check on any/all the above simply because they're the only ones available for site visits.

4. Find out the "next steps", in terms of what will happen to the information which has been gathered, and what will happen (if anything) with the complaint which was made. Many times the complaint will end right then and there, assuming that you don't have anything massively out of order with your operation. A typical response will be "a complaint was filed due to concerns about X, but we don't see any evidence of X. We'll be reporting back to the person who made the complaint to let them know that everything looks OK." And that might be the end of it.

5. If something is found which violates local, state or federal law, the vast majority of public agencies will give you a chance to remedy the situation. In those instances, they will give you very specific written information about what condition/situation was in violation, the laws which apply, the corrective action which is required, and the time interval during which those corrections need to be made.

   Suppose for instance that a neighbor objected to a large pile of rabbit manure in the backyard, out of concerns it would either be a disease risk, a pollution risk, and/or an attractant for rodents (any/all of which is possible). You would probably be informed that you'd need to dispose of that pile by such-and-such date, and that a follow up inspection or report would be required to document that the corrective action was taken. If you carry out the corrective action, you're done. If you don't, then there will be further complications depending upon the regulations involved. Those "complications" could be financial penalties, forfeiture of your animals, or even jail time if things really get out of hand. For the vast majority of situations, simply taking care of the problem, will take care of the problem.

6. Most agencies will also require that you make efforts to avoid this same situation from becoming a problem again in the future. So for instance if manure built up enough to cause a complaint once, be prepared not only to clean it up, but also take steps to show how you'll ensure it won't build up again.

Note: Sometimes the person who made the complaint, particularly if it was a neighbor, may want to participate in the site visit. While you are legally compelled to allow lawful inspection of your property by public servants/employees, you are NOT under any obligation to allow a private citizen access to your property. In other words, if the neighbor raises a stink about your rabbits and calls Animal Control to report you, you may have to submit to an Animal Control inspection but your neighbor doesn't have the right to be there alongside the AC officer. The neighbor in question can just wait at home for the officer's report.

I also wanted to note that while public agencies are generally held to certain ethical requirements and privacy laws, the same cannot be said for another enforcement group, namely a homeowner's association. Some neighborhoods and many subdivisions have privately organized, managed and enforced rules for what home owners or tenants may or may not do within the boundaries of those neighborhoods. They are not bound by the same rules that public agencies follow. Frankly, we've heard bigger horror tales from people who wanted rabbits and were prevented from having them due to homeowner association rules, than because of any local, state or federal regulation. If you live in a neighborhood or subdivision governed by one of these groups, check with them first to see what you are or are not allowed to do concerning rabbits. Even if you don't agree with the rules, you'll be in a position to make informed decisions once you know what the rules are.

Sale of Animal Products
I'm not going to belabor this point, in part because it is well outside of this writing's scope for discussing colony management. However, I did want to at least address the fact that different areas will have different laws for the sale of rabbit meat, and sometimes rabbit pelts or furs. If that is any part of your plans, it is in your best interests to at least determine what those local rules are, and what the penalties are if you don't follow them. It might be a slap on the wrist, or it might be substantial financial penalties and/or even jail time. This is an area where proactive education on your part will only help you make the best possible decisions for your particular situation.

Relations With Neighbors
Hard on the heels of the above topic, we come to the subject of keeping the neighbors happy. Since they are the ones who will be living in close proximity to your operation, and seeing, hearing, and/or smelling it all day every day, they are the most likely to eventually have some objections. As we saw in the Regulations chapter, everyone has different opinions on what is or is not "good animal husbandry". However, unlike passers-by who may not be around very often, your neighbors can either become a wonderful ally, or your worst nightmare, depending on how carefully you fold their concerns into your operation.

In the case of rabbits, noise isn't usually going to be an issue. However, smell might be if you have a lot of rabbits. While rabbit manure is good for the garden in reasonable amounts, a big messy pile of manure against a neighbor's fence is a good way to force a confrontation with said neighbor. Similarly, if your rabbits regularly tunnel under your fence and into the neighbor's flower patch, that's a problem. If their dog is absolutely fascinated with your rabbits and insists on either barking non-stop at the rabbits, or pacing up and down the fence line whenever your hops are out in the yard, that's a good time to consider mutually beneficial changes to your operation.

Our recommendation is to start off your operations with a healthy concern for how your animals might impact your neighbors, either now or in the future. And then work at finding ways to minimize that impact. In this instance, some might argue that you're under no legal obligation to do so. However, sheer common sense and even a thimbleful of community spirit would say that preserving

good neighborly relations is in everyone's best interest. So here are some specific recommendations:
1. If your neighbors have dogs, assume the dogs will be very interested in your rabbits. Try to minimize that interest, by not letting the rabbits go right up next to the fence, not piling up manures right next to the fence, and making sure your rabbits can't get into the neighbor's yard. If your rabbits cause the neighbor's dog to bark when either is outside, try to proactively work with your neighbor to ensure that the dog and the rabbits aren't outside at the same time.

   Alternately, set up the rabbit pen such that the rabbits aren't visible to the dog when they are outside. If the rabbits live outside 24/7, try to put up a visual barrier between them and the dog so that the dog can't see them moving around. And try to ensure that any manures and urine are either deposited into litter boxes or fresh bedding, and that soiled bedding is composted or otherwise used as quickly as possible so that smells don't build up. If there is no fence between your two yards, consider putting one up so that both the dog and the rabbits stay where they belong. The old adage that "Good fences make good neighbors" is as true now as it was when that saying first came to be.

2. If your neighbors have either dogs or cats, treat those animals as being potential predators for your rabbits. Review our suggestions for protection from predators, and use as many of them as you can. Only one or two measures will often fail, but using as many as possible

will make most rabbit operations far too much effort to mess with for anything other than a starving predator. If you are surrounded by cats and dogs on all sides, and/or one or more neighbors are simply not interested in any cooperative effort to keep their animals home, resign yourself to keeping your rabbits indoors, and/or in a very well protected outdoor pen. A sunroom might even be the best choice to ensure that there aren't any unfortunate accidental meetings.

3. If your neighbors are gardeners, offer them free rabbit manure as a way to sweeten their disposition towards your rabbits. Free fertilizer can inspire gardeners to tolerate a lot that they wouldn't otherwise put up with. If you are raising your rabbits for meat, you may also want to offer to share meat with your neighbors as a similar goodwill gesture.

Conclusion

As I indicated in the Introduction, this write-up of our experiences was something of a spontaneous idea. Two months ago, I didn't know I'd be staying up late or getting up early (ahem, earlier) to work on yet another writing project. Yet once the idea came to mind, it absolutely refused to leave me alone until I finally caved in and relented. My author self is still perturbed that some parts of this work will seem "thrown together" in almost shorthand, but that is how the information is apparently stored in my head. A series of notes-to-self which, taken together, form a record for what we've learned along the way. Frankly I'm not sure if that has been appealing or annoying for you the reader. Whichever the case, I hope that you were able to make use of at least something here.

As you have already seen, we are certainly not "arrived" yet at a perfect solution to either colony management or pasture management for rabbits. We'll probably be refining our ideas for years to come. One of my chief joys in life is to share what we've learned, and then learn from others who may have taken a completely different approach to the same problem, the same situation or the same goal. I figure there's more than one way to do something right, and maybe we can all learn from each other. If nothing else, hopefully this will fire up the creative juices for your own situation and you'll come up with something entirely new and wonderful. If that's the case, share! We'd love to hear from you on what worked, and what didn't.

In closing, I would wish for you a lot of successes with your rabbits. By success, I don't necessarily mean that you'll get it right the first time around; there is much opportunity to learn from mistakes. I also wish you a lot of joy with your rabbits, which I think can come even when we're making mistakes. They are such gentle, easy-going, forgiving creatures; any time spent with them is quality time. And the more we can give them license to truly act like rabbits, I think the better off everyone will be.

Photo Gallery

This is one of our early pens, where we tried to put the rabbits on pasture to graze, but then realized that the grass wouldn't come up through the wire flooring. So we switched mid-year to a deep bedding system by putting their feed hay directly on the ground each morning. You can see the day's hay supply in the lower left hand corner of the image. This is a single litter of rabbits, just prior to being sold. They're about 4 months old. Note that this cage system used 1"x2" rabbit wire for the sides, a 2"x2" panel for the roof. The blue and white tarps in the background were for rain protection, which is rare in our summertime. The bent PVC hoops arching over the cages would keep the tarps off the roof and sides, while the single upright PVC pole provided a "tentpole" point so that rainwater wouldn't pool on the tarp and cave in the rabbit pens.

Our first field shelter, with several different pen designs. The bachelor pen in the left back corner was 2" x2" fencing all around, and was terribly wobbly. The pens with the rounded roofs along the center, and in the front left corner, were our first weanling rabbit pens. They featured ½" x 1" sides and bottoms, with ½" hardware cloth enclosing each end. The entire frame was reinforced with concrete mesh, a heavy-gauge wire mesh with 4" x 4" openings. Those pens were overkill in terms of weight per unit size, and the hardware cloth wore out really fast along the bottom edge. The pens along the right side of the shelter were 1" x 2" rabbit wire for the sides, ½" x 1" rabbit wire flooring for the bottom, and 2" x 2" wire panels for the top. We liked those pens quite a bit, except the larger openings along the top allowed rodents into the pens. The rodents eventually learned to eat the alfalfa pellets.

One of our pre-colony cage setups. These homemade cages are made entirely from 1"x2" wire for the sides and roof, and ½" by 1" flooring wire for the floors. We did not have our rabbits out on pasture at this stage so there wasn't any concern about trying to get the grass to grow through the bottoms. We did however already feed hay at this point. Note the hay channel running down the middle. That allowed us to feed hay to each individual cage, without it being wasted due to trampling. That feature worked really well. While we eventually abandoned the individual cages, we converted them to pen use by removing the side cage panels to open up the cages to each other. We then cut large openings in the hay channels and framed off the sides of those channels, so that we could keep the hay channels while allowing passage back and forth from one side of the pen to the other. That setup worked really well and we continue to use it to this day for weanling pens. Note the shelter frame is supported by metal poles, and is much sturdier than our first-ever pasture shelter. Even this shelter, though, required us to shovel snow off the roofs in winter.

 Our first field shelter, made of PVC rods supporting a large (20' x 20') tarp. The PVC rod structure didn't hold up very well over the course of the summer, and we eventually abandoned that shelter design. Looking back now, it never would have survived our wintertime snows or windstorms. Also note the fencing surrounding the shelter. The plan was to remove the fencing, move each interior pen, then move the shelter, across the pasture about once every few days. That plan turned out to be totally unworkable for a single person (me) doing all that take-down/move/put-back-up work, while trying to get other things done on our operation. It was our first attempt at pasturing our rabbits, and we learned a lot about what didn't work from that experiment. I should note that the fencing surrounding the shelter didn't do much to keep predators or vermin out, or the rabbits in. The openings were simply too large and the grass was too thick such that the fence shorted out. Now we use 2"x4" welded or woven field fencing, depending on the enclosure, sometimes with hardware cloth along the base to keep the smaller rabbits in and the rodents out. We also use a few strands of hotwire now, at top and/or bottom of the fence, rather than the netting shown in this picture.

Above Left: A close up of one of our bachelor pens, using the 2" x 2" woven wire as the top, side and bottom portions of the pen. This fencing material was good to work with and held up well in terms of rust resistance. However, it was too floppy to use by itself as a pen material. Ideally it needs to be nailed or fastened to a rigid frame if it's going to be moved in any way. Also, the opening size allowed rodents to come and go freely, and small rabbits could squeeze through it.

Above Right: Another close up of the 2" x 2" bachelor pen, showing the floor. Note that most of the grass was matted down very quickly under the wire where the rabbits couldn't really reach it. That, combined with the fact that rodents could easily come up through it, and the rabbits' legs could fall through it during moves, caused us to abandon 2" x 2" mesh as flooring material.

A close up of two of the pen designs from our first pasturing experiment. We've already discussed the bachelor pen on the upper left portion of the picture. The much heavier ½" x 1" pen in the lower right is now visible. Note the hatch on the side which allowed access to the interior. We abandoned this half-round pen design because it didn't give the rabbits much headroom after they'd grown a few weeks. It was also much too heavy. But it did have the merits of making absolutely sure that the young weanling rabbits couldn't get out, and rodents couldn't get in.

Two close-ups of one of our more recent pen designs. Note the switch to the 1" x 2" rabbit wire for the sides and floor. This has given us the best combination of stiffness, light weight, and vermin control of all the different fencing panels we've experimented with. Note however that the grass was matted down in both pens.

One of our breeding pens, using a purebred New Zealand doe, a Californian buck, and a few of the doe's hybrid daughters from a different Californian buck. While this pen had its issues (note the floppy, saggy roof), it did give the rabbits plenty of room to hop around, eat their hay, socialize and generally be rabbits.

Our most recent design, out in the garden beds. We used 1"x 2" rabbit wire for the sides and top, with ½" x 1" standard rabbit flooring wire for the floor. The pen is roughly 48" wide by 72" long. Note that for shelter we used ½" PVC hoops to hold up a 10'x12' white tarp, which clips onto the end PVC hoops with specialty clips made for garden hoop houses. Each clip is sized for that particular diameter PVC hoop size. This pen very comfortably housed 6 rabbits all through the wettest winter we've had in 50 years. The carrier in the middle of the pen gives the rabbits a place to sleep and hide, although they don't use it very often. This was not a breeding pen so we didn't need to provide nest boxes, but the pen was large enough that we could have. The roof stayed solid during wind storms, and shed frequent snowfall quite easily. With this design, the deep bedding system inside stayed nice and dry. These rabbits are about to move out for the spring, and their deep bedding will be turned into the garden soil immediately underneath. Then the pen will move to another spot in the garden to start again. Of all our pen designs, this is the one we like the most (so far).

About the Author

Ms. Kerby spent much of her teenage years on horseback, haunting the old ranchlands in then-rural Douglas County, Colorado. After high school she attended the University of Colorado-Boulder where she earned her Bachelor's of Science in Environmental Biology, with an additional focus on technical writing and scientific journalism. Between 1988 and 2011, she held a variety of jobs where either her scientific and/or journalistic skills were exercised regularly. During that time she also lived in six different states, from one coast to the other and several points between. In 2000 she and her husband purchased an old rundown property and dilapidated farmhouse in western Washington state, and started the hard work of restoring the property to full production. That work is still underway. They added their first rabbit herd in 2001 and have had rabbits ever since. Farming and farm-related products became her sole source of income in 2011. As of this writing in 2017, the farm has become a very well diversified livestock operation, with hay, field crops, forestry and garden also competing for attention. Visit the farm's website at www.frogchorusfarm.com. Her favorite things besides farming, gardening and working with her animals, are drinking a hot cup of tea, reading an old book, listening to various types of music, or watching either westerns or sci-fi movies.

www.ingramcontent.com/pod-product-compliance
Lightning Source LLC
Chambersburg PA
CBHW052053070526
44584CB00017B/2153